William Roy

Experiments and Observations Made in Britain,

in Order to Obtain a Rule for Measuring Heights with the Barometer. By Colonel

William Roy. F. R. S.

William Roy

Experiments and Observations Made in Britain,
in Order to Obtain a Rule for Measuring Heights with the Barometer. By Colonel William Roy.
F. R. S.

ISBN/EAN: 9783337382155

Printed in Europe, USA, Canada, Australia, Japan

Cover: Foto ©berggeist007 / pixelio.de

More available books at **www.hansebooks.com**

XXXIV. *Experiments and Observations made in* Britain, *in order to obtain a Rule for measuring Heights with the Barometer.* By Colonel William Roy. F. R. S.

Read June 12 and 19, and Nov. 6 and 13, 1777.

INTRODUCTION.

IN philofophical inquiries of every kind, where any point is to be afcertained by experiments, thefe cannot be repeated too often, nor varied too much, in order to obtain the truth: for even when the utmoft precaution hath been ufed, and the greateft pains have been taken, it rarely happens, that they agree fo exactly, as to leave no room for doubt. Were it poffible at all times, to have experiments made in circumftances perfectly fimilar, a confiderable degree of confiftency might naturally be expected among the refults, whereof the mean would determine the point in queftion; but different men, making ufe of different inftruments, have different modes of conducting their operations, each purfuing the tract that feems to him the moft likely to infure fuccefs. Hence it is that a variety

of

of refults arife, and that things fometimes appear contradictory, or at leaft prefent themfelves under new forms,
difficult at firft fight to be accounted for, and therefore
apt to miflead, till by a farther inveftigation of the matter, the true caufes are difcovered. Even irregularities of
this fort are worthy of being communicated, that others
may know what hath happened before, and what, in like
cafes, they may expect to meet with, in the courfe of
their future inquiries. Improvements of every kind ad-
vance by flow degrees; and it is not until things have
been viewed in every poffible light, that the errors,
even of our own experiments, are difcovered, the points
in queftion ultimately afcertained, and the branch of
philofophy depending upon them, gradually brought
nearer to perfection.

Ever fince the difcovery made by TORRICELLI, the
barometer hath been applied, by different perfons, in different countries, to the meafurement of vertical heights,
with more or lefs fuccefs, according to the more or lefs
perfect ftate of the inftruments ufed, and the particular
modes of calculation adopted, by the obfervers. But of
all thofe who have hitherto employed themfelves in this
way, none hath beftowed fo much time and pains, or
fucceeded fo well, as Mr. DE LUC, of Geneva, F. R. S. In
two quarto volumes, publifhed fome years fince, that
                                                                    gentleman

gentleman hath given us the hiftory of the barometer and thermometer, with a very curious and elaborate detail of many years experiments, made by him, chiefly on the mountain Saleve. It would be totally fuperfluous here to enter into any circumftantial account of the method he makes ufe of; fince that hath already been fo fully illuftrated by two Fellows of the Royal Society, who have at the fame time given formulæ and tables, adapted to the meafures of this country, (Phil. Tranf. for 1774, vol. LXV. N° xx. and xxx.) that nothing farther can be defired on that head.

It may neverthelefs be neceffary juft to call to remembrance that the rule, deduced from the obfervations on Saleve, confifts of three parts. 1ft, The equation for the expanfion of the quickfilver in the tube, from the effect of heat, whereby the heights of the columns, in the inferior and fuperior barometers, are conftantly reduced to what they would have been in the fixed temperature of $54^{\circ}\frac{1}{4}$ of FAHRENHEIT, independant of the preffure they refpectively fuftained. 2d, When the mean temperature of the column of air to be meafured, is $69^{\circ}.32$, as indicated by thermometers expofed to the Sun's rays at its extremities; then the difference of the common logarithms, of the equated heights of quickfilver in the

two

two barometers, gives the altitude intercepted between them, in toifes and thoufandth parts, reckoning the three figures to the right hand decimals, and the others integers, the index being neglected. This temperature of 69°.32, when the logarithmic differences give the real height without any equation, is reduced to 39°.74, the new zero of Mr. DE LUC's fcale, when his formula is adapted to Englifh fathoms and thoufandth parts, inftead of French toifes. And laftly, when the mean temperature of the air is above or below 39°.74, an equation, amounting to $\frac{21}{10000}$ parts of the logarithmic height for each degree of difference, is, in the firft cafe to be added to, and in the laft fubtracted from, that refult, in order to obtain the real altitude.

In Mr. DE LUC's book, the experiments for afcertaining the expanfion of the quickfilver, are not given in detail; neither are the particular temperatures of the barometers fpecified. The winter feafon was however chofen for the purpofe; one being left in a cold room, and the other in a clofet, heated as high as could conveniently be fuffered. The operation having been repeated feveral times without any effential difference in the refults, this general conclufion is drawn, that between the temperatures of melting ice and boiling water, the expanfion of

the

the quickſilver is exactly ſix French lines, or $.532875$ decimal parts of an Engliſh inch. But it is to be obſerved, that the barometer ſtood then at 28.77525; whereas, if it had ſtood at 30 inches, it would have been $.555556$, becauſe the expanſion is in proportion to the length of the column. Farther, the interval between the freezing and boiling points in all thermometers, varies with the height of the barometer, or weight of the atmoſphere; and it is the cuſtom in England to make thermometers when the barometer ſtands at 30 inches; that is to ſay, 1.225 or 13.8 French lines, higher than when Mr. DE LUC's boiling point was fixed: and ſince from his experiments it appears, that each line of additional height in the barometer, raiſes the boiling point $\frac{1}{1134}$th part of the interval between that and freezing, it follows that $\frac{180}{1134} = 0.158 \times 13.8 = 2°.2$, will denote the number of degrees, that Mr. DE LUC's boiling point is lower than that of Engliſh thermometers, which reduces it to 209.8 of FAHRENHEIT, and makes the interval between freezing and boiling only 177.8 degrees. Hence the expanſion $.555556$, formerly found, muſt be increaſed in the proportion of 177.8 to 180, which gives for the total $.5624297$ or $.56243$, on a difference of temperature of $180°$. Thus the expanſion for each degree, ſuppoſing it to be arithmetical, or uniformly the fame in all parts of the fcale, will be .00312461 [a].

Having now fhewn the expanfion of quickfilver in the tubes of barometers refulting from the Geneva obfervations, I fhall next proceed to give fome account of thofe I made for that purpofe. They derive their origin from my having very accidentally obferved, that a fmall degree of heat, and of fhort duration, fenfibly affected the length of the column in Mr. RAMSDEN's portable baro-meter, whereof a view is given in plate XVI. The principal parts of this inftrument are a fimple ftraight tube, fixed into a wooden ciftern, which, for the conveniency of carrying, is fhut with an ivory fcrew, and, that being removed, is open when in ufe. Fronting this aperture is diftinctly feen, the coincidence of the gage-mark, with a line on the rod of an ivory float, fwimming on the furface of the quickfilver, which is raifed or depreffed by a brafs fcrew at the bottom of the ciftern. From this, as a fixed point, the height of the column is readily mea-fured on the fcale attached to the frame, always to $\frac{1}{500}$

(a) This paper having lately been communicated to Mr. DE LUC, he hath informed me, that the difference of temperature in his experiments, amounted to about 31° of REAUMUR, or 72° of FAHRENHEIT, above freezing: wherefore, .00312461 × 72 = .225 nearly, will denote the rate of expanfion from which he deduced that for 180°; and within thefe limits, it will hereafter be ound to differ very little from the refult of the prefent experiments.

I   part

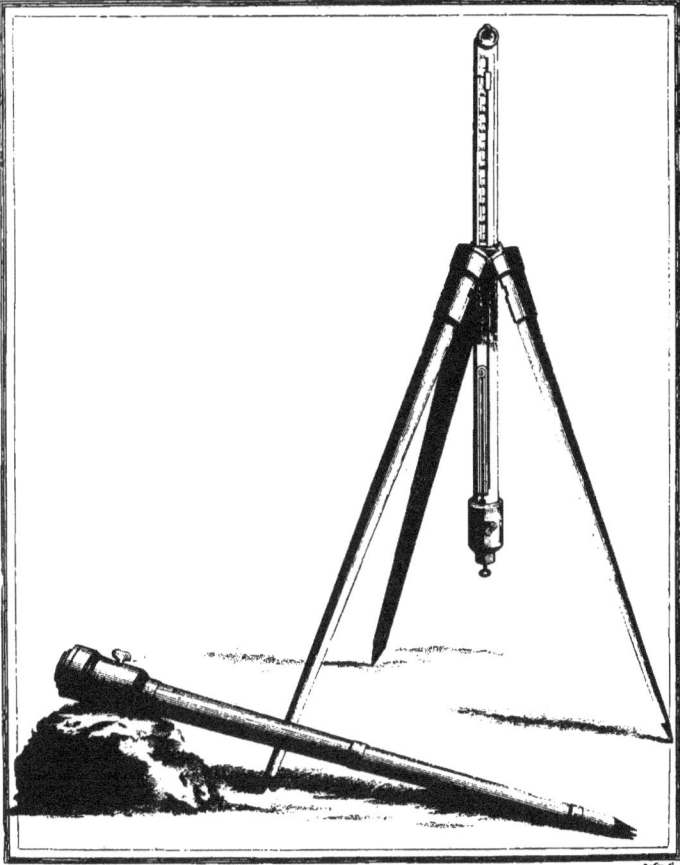

part of an inch, by means of a nonius moved with rack-work. A thermometer is placed near the ciitern, whofe ball heretofore, was ufually inclofed within the wood work, a defect that hath been fince remedied. The three-legged ftand, fupporting the inftrument when in ufe, ferves as a cafe for it, when inverted and carried from place to place. Two of thefe barometers, after the quick-filver in them hath been carefully boiled, being fuffered to remain long enough in the fame fituation, to acquire the fame temperature, ufually agree in height, or rarely differ from each other more than a few thoufandth parts of an inch, which were conftantly allowed for in calcu-lating altitudes, as well as in eftimating the rate of expan-fion, in the courfe of the following experiments.

SECTION I.

Experiments on the expanfion of quickfilver.

THE experiments made for this purpofe were nu-merous as well as various, and were therefore fubdivided into feveral claffes. To give a minute detail of them all, would be extremely tedious, and now wholly ufelefs, fince it was from thofe of the third clafs alone, that the

rate

rate as well as maximum of expanfion was afcertained: wherefore thofe of the two preceding claffes need only be mentioned in a general way.

The firft fet of the firft clafs comprehended fuch as were made with one barometer in a cold room, or in the open air, and the other in a room on the fame level with the former, where there was conftantly a fire, which was occafionally increafed, in order to augment the difference of temperature. When the heated barometer had remained feveral hours in an angle of the room, the difference of temperature of its quickfilver above that of the coldeft, as indicated by their refpective attached thermometers, rarely exceeded 10 or 12°, which, from a mean of many obfervations, gave an expanfion of .0333 decimals of an inch, for the 10° comprehended between 32 and 42° of FAHRENHEIT's thermometer. So far the refult arifing in this way, from fmall differences of temperature, will be found to agree with the third clafs of experiments.

But when, in the fecond fet of this firft clafs, the difference of temperature was augmented to 20 or 30°, by expofing the barometer within doors to a greater heat, or placing the fuperior one on the leads, whereby it received the direct and reflected rays of the Sun throughout the greateft part of the day, while the other was kept

in

in the cold area underneath, the rate of expansion for the first 10° exceeded that formerly found nearly in the proportion of three to two, while that for the second and third terms, of 10° each, diminished progressively.

The chief, though not the only cause of this great difference, as will appear hereafter, arose from the position of the ball of the thermometer, originally inclosed within the wood-work of the frame, which prevented it from receiving the heat so readily as the quickfilver in the tube; at the same time that it retained it longer, and consequently produced results in some degree fallacious.

Finding, from the first class of experiments, that much uncertainty remained with regard to the rate of expansion of quickfilver affected by these smaller degrees of heat, and that it was utterly impossible, from them, to determine its maximum for the 180° between freezing and boiling; I resolved to try, how much a column of 30 inches of quickfilver, carefully boiled in a tube, would lengthen, the same being placed with the open end upwards in a tin vessel, occasionally filled with pounded ice and water, and afterwards brought to boil, by means of a charcoal fire placed underneath? In this second class, it was easy to see, that the expansion of the tube containing the quickfilver, was necessarily to be taken into the account, and added to that apparently found by experi-

2 ment.

ment. This was of courſe to be done, either by ſuch differences as I could diſcern and meaſure, or by thoſe that had reſulted from the experience of others.

The nature of the apparatus, employed in this claſs of experiments, will be eaſily underſtood from plate XVII. where it is repreſented, as it was uſed in thoſe of the third claſs. In its firſt ſtate it was not quite ſo long, and a chafing-diſh with a charcoal fire, occupied the place of the ciſtern holding the quickſilver below. By means of a circular bit of tin, ſoldered edgeways in the center of the bottom, and an aperture in the middle of the lid, the tube was kept ſteadily in the axis of the veſſel. Other openings in the lid, ſerved for the admiſſion of the thermometer, and the application of a deal rod cloſe to the ſide of the tube, when its height was to be meaſured. The longitudinal expanſion of the glaſs was marked by a ſcratch thereon with a fine edged file at the top of the deal rod, when reſpectively at the temperatures of freezing and boiling. The apparent dilatation of the quickſilver was in like manner marked, by the coincidence of its ſurface with the lower edge of a braſs ring embracing the tube.

It having been found impoſſible to procure tubes whoſe bores were truly cylindrical, or of any uniform figure, the experiment was repeated, as often as poſſible,

in

in both ends of the fame tube, that the mean might be taken. But it frequently happened that the tube, which had undergone one or more experiments in one end, broke before any could be made with it in the other. In this cafe, the rate of expanfion in the laft end was taken from that given by fuch another tube, where it had fucceeded in both. The mean of five refults with the beft tubes, taken in this way, gave .4901 for the apparent expanfion of 30 inches of quickfilver, on 180° of FAHRENHEIT, between freezing and boiling, which being augmented by the apparent longitudinal dilatation of the glafs .0356 × 3 = .1068, the real expanfion is .5969; exceeding Mr. DE LUC's by more than $\frac{3}{100}$ths of an inch. If, however, Mr. SMEATON's dilatation of glafs, (.025 × 3 = .075) be fubftituted, inftead of that refulting from thefe experiments, the real expanfion of 30 inches of quickfilver will be .5651, which does not exceed it quite $\frac{1}{1000}$ parts of an inch.

In this clafs of experiments, having attended as diligently as poffible to all the circumftances, it feemed to me, that tubes with a fmall bore, and whofe glafs was thick, lengthened more than thofe, which had a larger bore and whofe glafs was thin: whence I was led to fuppofe, that folid glafs rods would dilate more in proportion, and confequently, fhew a ftill more perceptible difference.

difference. With the view of afcertaining this point, I procured four glafs rods near three feet long each, and of different diameters, the largeft being of the fize of the little finger, and the fmalleft about the thicknefs of a quill. One end of each, was fomewhat larger than the other, and was made perfectly fmooth, as that on which they were to reft when feverally meafured with the deal rod. They were then all placed in the tin veffel, in fuch a·manner, as to admit pounded ice rammed very clofe around them, and the interftices to be filled with water. Having remained in that ftate a full half hour, they were feverally meafured with the deal rod, whofe length of 32¼ inches was fcratched on each with the fharp edge of the file. This being done, the ice thrown out, and the veffel carefully wafhed, all the rods were replaced in it, immerfed in water, which afterwards was brought to boil. The fire being kept up, and the ebullition rendered as violent as poffible for half an hour, the glafs rods were then feverally meafured, by applying them one after another to the deal rod, ftanding with them in the boiling water. The experiment was repeated three times, on as many different days, without its being poffible to difcern, that any of the glafs rods had dilated more than that of deal, from a difference of temperature of 180°. In all of them, the freezing mark feemed accurately to coincide with the top

of

of the deal rod; whereas the dilatation of the tubes, by the ſame degree of heat, was always very viſible[b].

Finding from the ſecond claſs of experiments, whereof the general reſult hath now been given, that glaſs rods ſeemed not to lengthen more than deal; and that tubes of different bores, and probably too of different ſorts of glaſs, were ſuſceptible of different degrees of extenſion, which rendered it impoſſible, by this means, to aſcertain the longitudinal expanſion of the quickſilver they contained; I thought it neceſſary to recur once more to the barometer, and to try whether it could not be ſo contrived as to act in water of different temperatures, from freezing to boiling. This led me to the experiments of the third claſs: and in order to comprehend them thoroughly, it ſeems neceſſary to point out ſome few alterations which the apparatus underwent.

The center of the bottom being pierced on purpoſe, a braſs ſocket was prepared for it, wherein a hole was bored conically, to receive the ground-end of a barometer tube, of the ordinary length of $33\frac{1}{2}$ inches; the tube having been firſt ground in a ſeparate piece of braſs, and ulti-

---

(b) Since theſe experiments were made, the relative expanſion of 18 inches of one of the tubes and one of the rods that had been formerly employed, was found to be, by Mr. CUMMING's pyrometer, nearly as 4 to 1, from a heat approaching to that of boiling oil.

mately

mately in the focket itfelf, fitted it fo exactly, as to fuffer no water to pafs. The focket being inferted into the aperture at the bottom of the veffel, was firmly foldered to it for the reception of the tube, which was fo ground as to reach a full inch and a half below the furface of the brafs. It could not defcend farther, the ground parts in both being of the figure of the fruftum of an inverted cone. From the view in the plate it will appear, that underneath the veffel, a feparate ftand was placed, in order to fupport the iron ciftern containing the quickfilver. The diameter of the ciftern was fuch, that its ftand being occafionally moved, fo as to bring one fide of it clofe to the ground part of the tube, the other fide projected beyond the bottom of the veffel; and confequently permitted the rod of a float, refting on the furface of the quickfilver, to rife freely and parallel to the axis of the tube. The rod was of deal, $\frac{1}{16}$th of an inch fquare, carrying on its top a fcale, whofe zero lay in the lower furface of the float, and whereof the fix uppermoft inches, from 28 to 34, were divided into 20ths.

That the whole column of quickfilver might alternately be covered with the freezing mixture and water of different temperatures, and yet permit its furface to be feen, two eyes of plate glafs were fcrewed into fockets, foldered for that purpofe oppofite to each other, near the

top

top of the veffel, which, in the firft fet of the third clafs of experiments, was little more than 29 inches high. The top of the tube paffing through the aperture in the lid, one and a half or two inches of the vacuum generally rofe above the veffel. That the expanfion of the column might be meafured as nearly as poffible in that part of the tube fronting the center of the eyes, more or lefs quickfilver, according to the ftate of the atmofphere, was occafionally put into the ciftern, to raife or deprefs the furface of the column to the proper height. A thin brafs ring, whofe lower parts were made to fpring, embraced with fufficient force the upper part of the tube, permitting it at the fame time to be moved freely with the hand. It carried along with it a nonius index, projecting as far as the center of the rod, and confequently applying itfelf to the divifions of the fcale, which was kept in its proper pofition by paffing through a flit fitted for it in an arm attached to the lid. The divifions on the nonius being the fame with thofe of the barometer formerly defcribed, the height of the quickfilver could always be read off to $\frac{1}{500}$th part of an inch.

The quickfilver having been carefully boiled, as on former occafions, in the tube; and that being filled completely, and held with its open end upwards; the tin veffel was inverted over it, and lowered gradually, till the

ground

ground end could be inferted into the focket with fuch a
degree of force as to prevent it from being too eafily removed. The finger being then applied clofely to the
open end of the tube, the whole apparatus was turned
up, and placed over the ciftern into which the quickfilver
had previoufly been put, great care being taken not to
remove the finger till the lower extremity of the tube
was fairly immerfed into the quickfilver; when that in
the tube was permitted to defcend into an equilibrium
with the atmofphere. In the firft experiment it was
found that the water iffued by the eyes, and running
down the fide of the veffel, fell into the ciftern. In order
to remedy this inconveniency, a circular piece of tin was
foldered round the upper part of it, immediately below the
eyes; and a flat fpout, projecting from it, ferved as a gut-
ter to throw off the water from the ciftern, and from the
lamps made ufe of to bring that in the veffel to boil. Six
lamps, each with a double light, were fufpended around
the trunk of the veffel, to heat the water as equally as
poffible; though any irregularity of this kind was fufficiently guarded againft, by conftantly mixing it during the
operation. Another lamp of the fame kind ftood under
the ciftern, whereby the quickfilver there was kept at
the temperature of the water in the veffel, each having
its proper thermometer: this laft lamp was placed
                                                    and

and difplaced frequently, during the courfe of every experiment; for the heat was very expeditioufly communicated to the iron ciftern, and thence to the quickfilver
it contained ; and both were found to cool very faft, after
the lamp was removed. Such was the ftate of the apparatus, when the firft fet of this third clafs of experiments
was made. In thofe of the fecond fet, its height was farther augmented by tin foldered to the top, that a tube of
the ordinary length might be wholly immerfed in boiling water. The third and laft alteration confifted in the
occafional application of a detached tin cafe, equal in
diameter to the upper part of the veffel, having a hole in
its bottom to admit the top of a long tube to pafs. This
cafe was fo contrived, that its bottom ftood two inches
and a half higher than the lid of the veffel, thereby allowing room for the hand to move the index up or
down. In this ftate the apparatus is reprefented in the
view; and its various ufes will be beft underftood from
the account of the experiments, which were fubdivided
into four fets.

Thofe of the firft fet were made with tubes of a large
bore, upwards of three-tenths of an inch in diameter, of
the ordinary length, with a vacuum over the quickfilver
of two inches and a half or three inches, part of which
reached above the top of the veffel. The mean of three

experi-

experiments gave .5258, for the total dilatation of 30
inches of quickfilver, on 180° between freezing and boiling; that, anfwering to the firft 20°, between 32° and
52°, was .0688; that, for the 20° in the middle of the
fcale, between 112° and 132°, was .058; and the rate for
the laft 20°, between 192° and 212°, was only .041.
From this firft fet of the third clafs of experiments, it appeared evident, that the expanfion of 30 inches of quickfilver in the barometer, fuffering a heat equal to 180° of
FAHRENHEIT, inftead of exceeding Mr. DE LUC's, as appeared to be the cafe from the refults of the open tube,
really fell fhort of it: and inftead of being arithmetical or
uniformly the fame, for equal changes of temperature, was
actually progreffive; the expanfion anfwering to the lower
part of the fcale, being greater than that correfponding to
the middle; which again exceeded that for high temperatures. In thefe experiments, when the water had acquired a heat 20 or 30 degrees greater than that of the open
air, a certain duftinefs was perceived in the vacuum of
the tube. At 100° of FAHRENHEIT, or thereabout, this
appearance had fo far increafed, as to fhew clearly, that
it could proceed from no other caufe than a vapour arifing from the furface of the heated quickfilver, quite in-
vifible, till, by its condenfation in the cold part of the
tube, it was formed into balls, every where adhering to

its

its fides and fummit. Thefe globules were very fmall near the furface of the water, augmenting gradually as they approached the top of the tube, where they were greateft: their bulk increafed with the heat; and when the water was at or near boiling, they would fometimes unite, and defcend by their own gravity, along the fides of the tube, into the general mafs. Hence the progreffive diminution of the rate of expanfion of the column of quickfilver in the barometer, perceptible even in the firft clafs of experiments, is eafily accounted for by the refiftance of the elaftic vapour[c], acting againft the top of the tube, which was here colder than the reft.

But in the application of the barometer to the meafurement of heights, the whole inftrument is of the fame temperature; wherefore, in the fecond fet of this third clafs of experiments, the tin veffel was heightened, that tubes of the ordinary length, placed in it, might be wholly immerfed in boiling water. The mean of four experiments, which agreed very nearly among themfelves, gave .5117 for the total expanfion between freezing and boiling; for the 20°, between 112° and 132°.059; and for the laft 20°, between 192° and 212°.046. In thefe

(c) Having mentioned the circumftances to Mr. RAMSDEN, it firft occurred to him, that the refiftance of the elaftic vapour was the caufe of the diminution in the rate of expanfion.

experi-

experiments; the tube being wholly covered with boiling water, no condenſation of vapour took place in the vacuum; and therefore no particles of quickſilver were ſeen adhering to the upper part of the tube. When the water boiled, the reſiſtance of the vapour was greater than in the preceding ſet, and the total expanſion leſs. Theſe two reſults ſerve ſtrongly to confirm each other: it is, however, the laſt that furniſhes the data for conſtructing the table of equation depending upon the heat of the quickſilver in the barometer, of which table we ſhall give an account hereafter.

Finding, from the compariſon of theſe two ſets of experiments with each other, that the maximum and rate of expanſion ſeemed to vary with the length of the vacuum above the quickſilver, I was adviſed to try[d] what might be the reſult, when the vacuum was much longer than in the common barometer.

The third ſet of experiments of this claſs was therefore made with a tube ſomewhat narrower in the bore than the former, and whoſe vacuum was $14\frac{1}{2}$ inches in length, whereof $11\frac{1}{2}$ reached above the top of the veſſel. The mean of three obſervations gave .5443 for the total expanſion on 180°; that for the firſt 20° was .067; for

(d) Dr. BLAGDEN, who afterwards aſſiſted in ſome of the firſt experiments with the manometer, propoſed that with the long tube.

the

the 20° in the middle of the fcale .058; and for the uppermoft 20°.065 : whence the mean rate for every 20°,
is nearly .0605[r]. In this fet, the condenfation in the vacuum of the tube was particularly attended to: it began,
as in thofe of the firft fet, immediately above the furface
of the boiling water, which was always kept an inch or
two above the top of the column : the lowermoft globules were very fmall, increafing gradually till they got
without the lid of the veffel, where they were the largeft;
thence they diminifhed uniformly upwards, and difappeared entirely three or four inches below the top of the
tube. Though the rate for the middlemoft 20°, in thefe
laft experiments, be below the mean, probably from fome
inaccuracy in obfervation; yet, being compared with the
former fets, they ftill ferve to corroborate each other: for
in thefe with the long tube, the vacuum feems to have
been either completely maintained, or nearly fo; and we
accordingly find the maximum of expanfion increafed,
and its rate rendered nearly uniform, as will be farther
confirmed from what follows.

(r) Mr. CAVENDISH, who affifted in the firft part of the experiments with
the open tube, informed me, that, in thofe made by his father Lord CHARLES,
the difference between the expanfion of quickfilver and glafs, from 180° of heat,
was .469. If to this we add Mr. SMEATON's dilatation of glafs, the total
expanfion of 30 inches of quickfilver will be .544, which agrees with the
experiments in the long tube, and gives a rate of only .003022 for each degree.

I have

I have already had occafion to mention that a detached tin cafe was fometimes applied above the veffel, in which ftate it is reprefented in the view. This method was thought of during the operations with the long tube, in order to try whether the vacuum was completely maintained by the temperature of the open air? For this purpofe the cafe was placed on the ftones of the yard, with a fmall tube inferted in it, to preferve an open paffage in the middle: it was then rammed full of a compofition of falt and ice; and afterwards fixed on the top of the long tube. The degree of cold thus applied round the greateft part of the vacuum, muft have been very great, probably near the zero of FAHRENHEIT; yet it produced no vifible alteration in the height of the column of quickfilver, which ftill remained in boiling water below, and fhould have rifen, if the vacuum had been formerly incompleat. As it would have occafioned much trouble to have lengthened the feveral parts of the apparatus fo as to have kept the long tube wholly in boiling water, the counterpart of this laft experiment was not made in the accurate manner it ought: neverthelefs, the tin cafe, being emptied of its cold compofition, was placed on the tube as before, and filled with boiling water; which, joining with the intermediate fteam arifing from that in the veffel below, muft have kept the whole nearly in the

fame

fame temperature. The confequence of this application was, that the column fhortened about $\frac{2}{100}$ths of an inch; which feems to prove, that the quickfilver vapour now reached the fummit of the tube, and, acting againft it, overcame, by fo much, the preffure of the atmofphere.

I fhould now proceed to give fome account of the fourth fet of this laft clafs of experiments, made on the con-denfation of the quickfilver, by means of artificial cold, below the temperature of the air. Previoufly however to this, it may not be improper to take notice, in a more general way, of fome others that were made on expan-fion; as tending, with certain circumftances yet to be mentioned, not only to confirm thofe already defcribed, but likewife to account for many irregularities that occur in operating with barometers.

In the courfe of the preceding experiments, from ac-cidents of various kinds, it was often neceffary to reboil the quickfilver; and in that operation, many tubes were broken. The frequent removal of the focket from the bottom of the veffel, in order to get others ground for it, became at laft very troublefome; and made more caution neceffary, in boiling fuch as were ground, efpecially in frofty weather, which happened to be the cafe in the laft days of March, 1775: wherefore it was thought beft in the interim to try, what might be the expanfion of a column

4 S 2   of

of quickfilver, carefully put into the tube, but not boiled
therein?

With this view, the ftandard barometer and apparatus
were left out during the night of the 29th, that they
might acquire the fame temperature, which was found
next morning to be 34°$\frac{1}{7}$; the unboiled quickfilver ftanding $\frac{1}{100}$th of an inch higher than that which had been
boiled. The lamps being applied to the veffel, the
lengthening of the unboiled column was perceived, on
the whole, to be more irregular, and the progreffive diminution quicker, than in former experiments; fo as to
give, for the maximum of expanfion, only .443 for 180°.

On the morning of the 31ft, the unboiled column,
which on the preceding day had been the higheft, was
lower than the other by near $\frac{1}{100}$ths of an inch, the temperature of both being 31°$\frac{1}{7}$. As the water acquired heat
from the application of the lamps, the rate of expanfion
diminifhed; and, at boiling, was only .405 for 180°. The
operation of the 30th feems to point out, in a manner
fufficiently conclufive, that the air contained in the unboiled quickfilver, rendered its fpecific gravity lefs, than
that which had been boiled even a great while before;
fince it required a longer column of the firft, to counter-
balance the weight of the atmofphere. And though the
vacua might poffibly, at the beginning, have been equally
                                            compleat

compleat in both; yet they could not continue long fo: for the air efcaping gradually from the unboiled quickfilver, its elafticity increafing with the heat, and uniting with the quickfilver vapour, muft have refifted the dilatation of the column, and rendered it lefs than on former occafions; which actually appeared from experiment. This is farther confirmed by the obfervations of the fubfequent day; for now the unboiled column was become the fhorteft, owing no doubt to more air having afcended, and rendered the vacuum ftill more incompleat. Thus, the caufes of refiftance increafing, the dilatation is leffened in a fuperior degree.

The other circumftances to be mentioned, occurred on the 12th of April. After finifhing one of the experiments of the fecond clafs, and when the water had cooled to 192°, the veffel, by accident, received a fudden jolt, whereby the mouth of the tube muft have been raifed, for a moment, out of the quickfilver in the ciftern. In a few minutes after this, intending to obferve how far the column had fhortened from the decreafing heat, I was furprized to find, that the quickfilver had wholly difappeared in the tube, and was funk fo low as not to be feen by looking obliquely down at the eye of the veffel. It was then certain that air, and probably a particle of moifture along with it, had afcended into the upper part of the tube, whereby

whereby the vacuum was deftroyed in fo remarkable a degree. Since this accident made it neceffary to reboil the quickfilver, the water (then between 180° and 190°) was let out by the cock fixed for that purpofe at the bottom of the veffel; but before it could be intirely drawn off, the tube and its contents, had fo fenfibly felt the condenfing force of the furrounding atmofphere, then about 48°, that the quickfilver had rifen again, and prefented itfelf oppofite to the eye of the veffel, fomething lower indeed than where it formerly ftood. On this difcovery, and as foon as water could be boiled for the purpofe, the veffel was filled again, when the quickfilver fubfided, as before, quite out of fight; and on drawing off the water a fecond time, it rofe anew, feemingly to its former height.

The appearance, which this accidental circumftance produced, was fuch, as naturally fuggefted that farther experiments might have been made, varied as much as poffible from each other, by the admiffion of different quantities of air, or of air and moifture intermixed. But the nature of the veffel rendering it impoffible to fee, and confequently to meafure, the motion of the quickfilver, occafioned by the alternate expanfion and condenfation of the elaftic vapour contained in the upper part of the tube, and which could not have been accomplifhed

plifhed without many troublefome alterations in the apparatus, therefore nothing of the kind was attempted. From the circumftances juft now mentioned, it will be readily conceived, how much care is neceffary in operating with barometers for the meafurement of heights, that the vacua be as nearly as poffible compleat; and particularly, that no moifture get up into the tube. I now proceed to the fourth and laft fet of experiments.

Having found, from the two firft fets of this clafs, the rate of expanfion of a column of quickfilver, in the tube of a barometer of the ordinary length, to be progreffive and not arithmetical; and that its maximum, for the 180° comprehended between feeezing and boiling, was lefs than had been fuppofed; I thought it proper to try, by means of artificial cold, whether the condenfation, for the 32° below freezing, followed nearly the fame law?

For this purpofe the tin veffel, containing the ground tube, was rammed quite full of pounded ice and falt, as well as the tin ftand holding the iron ciftern below. In this operation, twelve pounds of ice and four pounds of falt were employed, whereby the mean temperature of the mixture was reduced to + 4° of FAHRENHEIT. But before the eyes of the veffel could be fufficiently freed from the compofition, fo as to permit the furface of the column to be diftinctly feen and read off; it had rifen to

5   + 14°

+ 14°; the temperature of the air, and alſo of the ſtandard barometer, being at the ſame moment 49°$\frac{1}{7}$. The obſerved condenſation, ariſing from this difference of 35°$\frac{1}{7}$, was $\frac{12}{100}$ths of an inch; or .1189, when reduced for the height of the barometer, which then ſtood at 30.296. Hence the condenſation for 32° is .1072, or .00335 for each degree. In this day's experiment, when the temperature of the mixture had riſen to 32°, that of the air and ſtandard barometer was 52°$\frac{1}{7}$; whence the reduced difference, for the 20° between 32° and 52°, was found to be .0664, anſwerable to former experiments.

The ſame experiment was repeated two days after, with great care, the veſſel being filled no higher than the ſurface of the quickſilver. The mean temperature of the mixture was now +4°, and that of the ſtandard barometer 49°$\frac{1}{4}$. The obſerved condenſation, ariſing from this difference of 45°$\frac{1}{4}$, was $\frac{162}{1000}$; or .1594, when reduced for the height of the barometer, then ſtanding at 30,416: hence the rate for 32° is .1127, or .003,522 for each degree. When the temperature of the mixture had riſen to 32°, that of the air was 51°: whence the augmented rate for the 20°, between 32° and 52°, was found to be .0662.

From the mean of theſe two experiments it appears, that the condenſation of a column of 30 inches of
quickſilver

Baßer sc.

quickfilver in the barometer, affected by the 32° of cold below freezing, is .1099: and that the expanfion from 20° of heat, between 32° and 52°, is .0663, a number agreeing perfectly well with former refults. If the con-denfation .1099 thus found, be added to the expanfion .5117 arifing from the fecond clafs of experiments, we fhall have .6216 for the total difference of height of the columns of quickfilver in two barometers, fuftaining the fame preffure, but differing from each other in their tem-peratures 212° of FAHRENHEIT's thermometer.

The feries of numbers expreffed in the annexed table, agreeing in all effential refpects with the expanfions found by experiment, will therefore fhew that which correfponds to any intermediate temperature, for every 10° of the fcale.

Rate of expansion of a column of quicksilver in the tube
of a barometer.

	Temperature.	Expansions.	Differences.	2d Differences.
Expansion above 32° of FAHRENHEIT; equation to be subtracted from the height of the column of quicksilver of 30 inches.	212	.5117		
	202	.4888	.0229	
	192	.4652	.0236	
	182	.4409	.0243	.0007
	172	.4159	.0250	
	162	.3902	.0257	
	152	.3638	.0264	
	142	.3367	.0271	
	132	.3090	.0277	
	122	.2807	.0283	
	112	.2518	.0289	.0006
	102	.2223	.0295	
	92	.1922	.0301	
	82	.1615	.0307	
	72	.1302	.0313	
	62	.0984	.0318	
	52	.0661	.0323	
	42	.0333	.0328	
	32	.0000	.0333	.0005
Condensation below 32° of FAHRENHEIT; equation to be added.	22	.0338	.0338	
	12	.0681	.0343	
	2	.1029	.0348	
	0	.1099	.0070	

Con-

Conftruction and application of the table of equation, for the expanfion of the quickfilver in the tubes of barometers.

In the introduction to this paper there was occafion to remark, that in the application of the barometer to the meafurement of heights, various modes of calculation had been adopted. The eafieft and beft method feems however to be, by means of the tables of common logarithms, which were firft thought of by Mr. MARIOTTE, and afterwards applied by Dr. HALLEY, Mr. BOUGUER, Mr. DE LUC, and others. They have all proceeded on the fuppofition, that air is a truly homogeneous and elaftic fluid, whofe condenfations being proportionable to the weights with which it is loaded, its dilatations are in the inverfe of the weights; and in confequence of this law, that the heights of the atmofphere afcended, are in geometrical progreffion, while the correfponding fucceffive defcents of the quickfilver in the tube of the barometer, are in arithmetical progreffion.

Mr. DE LUC makes ufe of an arithmetical or uniform equation for the heat of the quickfilver in his barometers, whereby their relative heights are reduced to what they would have been in the fixed temperature of $54°\frac{1}{4}$

of

of ꜰᴀʜʀᴇɴʜᴇɪᴛ. In the formulæ adapting his rule to
Englifh meafures (Phil. Tranf. vol. LXVII. N° xx. and
xxx.) hath been fhewn, that the eafieft and fimpleft method is, to make the difference of temperature of the two
barometers the argument for the equation; and that it is
fufficient to reduce either column to what would have
been its height in the temperature of the other. But
whatever may heretofore have been the method of ufing
the equation for the heat of the quickfilver, while it was
confidered as arithmetical; now that it hath been fhewn,
from the preceding experiments, to be progreffive, there
feems at leaft to be propriety in applying to each barometer the equation anfwering to its particular temperature. And though, for this purpofe, any fixed temperature might have been affumed at pleafure, as that to
which both barometers were to be reduced; yet, the
freezing point being fundamental in all thermometers,
and that being likewife the zero of the fcale for the equation depending on the heat of the air, as will be fhewn
hereafter, it hath been preferred to any other.

From the experiments it appears, that a column of
quickfilver of the temperature of 32°, fuftained, by the
weight of the atmofphere, to the height of 30 inches in
the barometer, when gradually affected by different degrees of heat, fuffers a progreffive expanfion; and that,

having

having acquired the heat of boiling water, it is lengthened $\frac{5117}{10000}$ parts of an inch: alſo, that the ſame column, ſuffering a condenſation by 32° of cold, extending to the zero of FAHRENHEIT, is ſhortened $\frac{1099}{10000}$ parts, the weight of the atmoſphere remaining in both caſes unaltered; but that, in the application of the barometer to the meaſurement of altitudes, ſince the preſſure and length of the column change with every alteration of vertical height, the equation, depending on the difference of temperature of the quickſilver, will neceſſarily augment or diminiſh by a proportionable part of the whole. Thus, if the weight of the atmoſphere ſhould at any time be ſo great as to ſuſtain 31 inches of quickſilver, the equation for difference of temperature will be juſt $\frac{1}{15}$th part more than that for 30 inches; at 25 inches it will be $\frac{1}{6}$ths; at 20 inches $\frac{1}{3}$ds; at 15 inches $\frac{1}{1}$; and at 10 inches only $\frac{1}{3}$d of that deduced from experiment.

It is upon theſe principles that the annexed table of equation hath been conſtructed, for differences of temperature extending to 102° of the thermometer, and for heights of the barometer from 15 to 31 inches; beyond which limits, it is not probable, that many barometrical obſervations will be made. The firſt or left-hand column, ſhews the height of the barometer for every half inch, from 31 to 25; thence for every inch downwards

to

to 20; the 15th inch being half of the obferved expanfion. The five next columns towards the right, comprehend the additive equation for condenfations, anfwering
to 0°, 12°, and 22°, with their intermediate differences;
thofe that are progreffive, as arifing from difference of
temperature, being ranged horizontally; and thofe that
are arithmetical, as depending on the height of the barometer, being placed vertically. The temperature of
32° requires no equation, and the thirteen columns from
thence towards the right hand, contain the fubtractive
equations for expanfion, correfponding to every 10° as
far as 102°, with their progreffive and arithmetical differences ranged as before. By means of thefe differences,
the equation for intermediate temperatures may readily
be taken out by infpection. Hence is deduced the firft
part of the rule for meafuring heights by the barometer.
When the temperature of the quickfilver is below 32° of
FAHRENHEIT, add the correfponding equation for condenfation to the obferved heights of the columns refpectively; when above 32°, fubtract the equation for expanfion from the obferved heights of the columns refpectively; with which equated heights of quickfilver, expreffed in 1000th parts of an inch, the tables of logarithms are to be entered.

Table

Table, ſhewing the equation to be applied to the obſerved height of quickſilv
temperature extending to 102° of FAHRENHEIT: whereby the column is redu

Obſerved height of quickſil- ver in the Barom.	Condenſation below 32°; Equation to be added to the height of the quickſilver in the barometer.					32ⁿ	Expanſion above 32°; I			
	0°	Diff.	12°	Diff.	22°		42°	Diff.	52°	Diff.
31	.1136	.0432	.0704	.0355	.0349		.0344	.0339	.0683	.0334
30½	.1118	.0425	.0693	.0349	.0344		.0339	.0333	.0672	.0328
30	.1099	.0418	.0681	.0343	.0338		.0333	.0328	.0661	.0323
29½	.1081	.0411	.0670	.0338	.0332		.0327	.0323	.0650	.0318
29	.1063	.0405	.0658	.0331	.0327		.0322	.0317	.0639	.0312
28½	.1045	.0398	.0647	.0326	.0321		.0316	.0312	.0628	.0307
28	.1026	.0390	.0636	.0321	.0315		.0311	.0306	.0617	.0301
27½	.1008	.0384	.0624	.0314	.0310		.0305	.0301	.0606	.0296
27	.0990	.0377	.0613	.0309	.0304		.0300	.0295	.0595	.0291
26½	.0971	.0370	.0601	.0302	.0299		.0294	.0290	.0584	.0285
26	.0953	.0363	.0590	.0297	.0293		.0289	.0284	.0573	.0280
25½	.0935	.0356	.0579	.0292	.0287		.0283	.0279	.0562	.0275
25	.0916	.0349	.0567	.0285	.0282		.0278	.0273	.0551	.0269
24	.0880	.0336	.0544	.0273	.0271		.0266	.0263	.0529	.0258
23	.0843	.0321	.0522	.0263	.0259		.0255	.0252	.0507	.0248
22	.0897	.0408	.0399	.0251	.0248		.0244	.0241	.0485	.0237
21	.0770	.0294	.0476	.0239	.0237		.0233	.0230	.0463	.0226
20	.0733	.0280	.0453	.0228	.0225		.0221	.0220	.0441	.0215
15	.0550	.0211	.0339	.0170	.0169		.0166	.0165	.0331	.0161

Side annotations (braces): .00183, .00366, .00183 (0° column); .00114, .00228, .00114 (12°–22° columns); .00563, .001126, .00563 (22° column); .000555, .001110, .000555 (42° column); .001102, .002204, .001102 (52° column).

Vertical text in 32° column: When the temperature of the quickſilver in the barometer is 32° of FAHRENHEIT, no equation is neceſſary.

ickſilver in the barometer, from 15 to 31 inches; and for differences of
s reduced to the height it would have ſtood at in the temperature of 32°.

e 32°; Equation to be ſubtracted from the height of the quickſilver in the Barometer.

Diff.	62°	Diff.	72°	Diff.	82°	Diff.	92°	Diff.	102°
.0334	.1017	.0328	.1345	.0324	.1669	.0317	.1986	.0311	.2297
.0328	.1000	.0323	.1323	.0319	.1642	.0312	.1954	.0306	.2260
.0323	.0984	.0318	.1302	.0314	.1616	.0306	.1922	.0301	.2223
.0318	.0968	.0312	.1280	.0309	.1589	.0301	.1850	.0296	.2186
.0312	.0951	.0307	.1258	.0304	.1562	.0296	.1858	.0291	.2149
.0307	.0935	.0302	.1237	.0298	.1535	.0291	.1826	.0286	.2112
.0301	.0918	.0297	.1215	.0293	.1508	.0286	.1794	.0281	.2075
.0296	.0902	.0291	.1193	.0288	.1481	.0281	.1762	.0276	.2038
.0291	.0886	.0285	.1171	.0283	.1454	.0276	.1730	.0271	.2001
.0285	.0869	.0281	.1150	.0277	.1427	.0271	.1698	.0266	.1964
.0280	.0853	.0275	.1128	.0272	.1400	.0266	.1666	.0261	.1927
.0275	.0837	.0269	.1106	.0267	.1373	.0261	.1634	.0256	.1890
.0269	.0820	.0265	.1085	.0261	.1346	.0256	.1602	.0251	1853
.0258	.0787	.0254	.1041	.0252	.1293	.0245	.1538	.0240	.1778
.0248	.0755	.0243	.0998	.0241	.1239	.0235	.1474	.0230	.1704
.0237	.0722	.0232	.0954	.0231	.1185	.0225	.1410	.0220	.1630
.0226	.0689	.0222	.0911	.0220	.1131	.0215	.1346	.0210	.1556
.0215	.0656	.0211	.0867	.0210	.1077	.0204	.1281	.0201	.1482
.0161	.0492	.0158	.0650	.0158	.0808	.0153	.0961	.0150	.1111

SECTION

SECTION II.

Experiments on the expanfion of air in the Manometer.

WITH refpect to order of time, the manometrical experiments were made fubfequently to the chief part of the barometrical obfervations, from which alone an approximate rule had previoufly been deduced for the meafurement of heights: neverthelefs, in this paper it feemed to me beft, that what related to the expanfion of air in one inftrument, fhould immediately fucceed the expanfion of quiokfilver in the other.

The thermometer made ufe of in thefe experiments is above four feet long. Its fcale extends from − 4° to + 224° of FAHRENHEIT, each degree being more than $\frac{2}{10}$ths of an inch: when the barometer ftood at 30 inches, its boiling point was fixed in the tin veffel formerly defcribed. Mr. RAMSDEN's thermometers generally rife in the fame veffel 2 1 3°$\frac{1}{2}$; and the long thermometer, being placed in the veffel he makes ufe of to fix his boiling points, rifes only to 210°.

The manometers were of various lengths, from four to upwards of eight feet: they confifted of ftraight tubes,

5   whofe

whofe bores were commonly from $\frac{1}{13}$th to $\frac{1}{15}$th of an inch in diameter. The capacity of the tube was carefully meafured, by making a column of quickfilver, about three or four inches in length, move along it from one end to the other. Thefe fpaces were feverally marked, with a fine edged file, on the tubes; and transferred from them to long flips of pafte-board, for the fubfequent conftruction of the fcales refpectively belonging to each. The bulb, attached to one end of the manometer at the glafs houfe, was of the form of a pear, whofe point being occafionally opened, dry or moift air could be readily admitted, and the bulb fealed again, without any fenfible alteration in its capacity.

The air was confined by means of a column of quickfilver, long or fhort, and with the bulb downwards or upwards, according to the nature of the propofed experiment. Here it muft be obferved that, from the adhefion of the quickfilver to the tube, the inftrument will not act truly, except it be in a vertical pofition; and even then, it is neceffary to give it a fmall degree of motion, to bring the quickfilver into its true place; where it will remain in equilibrio, between the exterior preffure of the atmofphere on one fide, and the interior elaftic force of the confined air on the other.

All

All the experiments were made when the barometer was at, or near, 30 inches. When the bulb was downwards, the height of the barometer at the time of obfervation, augmented, and when upwards, diminifhed by the number of inches of quickfilver in the tube of the manometer, expreffed the denfity of the confined air.

Pounded ice and water were ufed to fix a freezing point on the tube; and by means of falt and ice, the air was farther condenfed, generally four, and fometimes five or fix degrees below zero. The thermometer and manometer were then placed in the tin veffel, among water which was brought into violent ebullition; where having remained a fufficient time, and motion being given to the manometer, a boiling point was marked thereon. After this the fire was removed, and the gradual defcents of the piece of quickfilver, correfponding to every 20 degrees of change of temperature in the thermometer, were fucceffively marked on a deal rod applied to the manometer. It is to be obferved, that both inftruments, while in the water, were in circumftances perfectly fimilar; that is to fay, the ball and bulb were at the bottom of the veffel.

In order to be certain that no air had efcaped by the fide of the quickfilver during the operation, the manometer was frequently placed a fecond time in melting

ice. If the barometer had not altered between the beginning and end of the experiment, the quickfilver always became ftationary at or near the firft mark. If any
fudden change had taken place in the weight of the atmofphere during that interval, the fame was noted, and
allowance made for it in afterwards proportioning the
fpaces.

Long tubes, with bores truly cylindrical or of any
uniform figure, are fcarcely ever met with. Such however as were ufed in thefe experiments, generally tapered
in a pretty regular manner from one end to the other.
When the bulb was downwards, and the tube narrowed
that way, the column of quickfilver confining the air
lengthened in the lower half of the fcale, and augmented the preffure above the mean. In the upper half,
the column being fhortened, the preffure was diminifhed
below the mean. In this cafe, the obferved fpaces both
ways from the center, were diminifhed in the inverfe
ratio of the heights of the barometer at each fpace, compared with its mean height. If the bore widened towards the bulb when downwards, the obferved fpaces,
each way from the center, were augmented in the fame
inverfe ratio; but in the experiments on air lefs denfe
than the atmofphere, the bulb being upwards, the fame
equation was applied with contrary figns: and if any

6   extra-

extraordinary irregularity took place in the tube, the correfponding fpaces were proportioned both ways from
that point, whether high or low, that anfwered to the
mean.

The obferved and equated manometrical fpaces being
thus laid down on the pafte-board containing the meafures of the tube; the 212° of the thermometer, in exact
proportion to the fections of the bore, were conftructed
along-fide of them: hence the coincidences with each
other were eafily feen; and the number of thermometrical degrees anfwering to each manometrical fpace,
readily transferred into a table prepared for the purpofe.

I have already had occafion to remark that, from the
operations of the barometer alone, an approximate rule,
or mean equation, had been obtained for the meafurement of heights; but as, among the refults, irregularities
were now and then met with, doubts naturally arofe,
whether the equation, inftead of being confidered as uniform, might not follow an increafing or diminifhing progreffion? Without an infinite number of obfervations,
in very different temperatures above and below the zero
of the fcale, this point could not poffibly be determined
by the barometer: wherefore the firft and chief thing
propofed to be difcovered by the manometrical experiments was, whether common air, occafionally rendered

more

more or lefs denfe, by the addition or fubtraction of
weight, expanded equally with quickfilver, when affected with the fame degrees of heat? According to the
ratio that took place between the expanfion of quickfilver and air, above and below zero, I intended to regulate
the barometrical equation already found, without re-
garding the proportion of the bulb to the bore of the
manometer; or in other words, without paying any attention to the actual expanfion of the air confined in that
inftrument.

But after a great number of thefe firft experiments
had been made, it was judged proper to compute the
actual expanfion of 1000 equal parts of air in the manometer, from a heat of 212°; wherefore, in the laft, the
accurate capacity of the bulb, with refpect to the bore,
was determined; at the fame time that the original mode
of comparing the thermometrical with the manometrical
fpaces, was ftill adhered to.

It is eafy to conceive in experiments of this very de-
licate nature, part of which, namely thofe on air lefs
denfe than the atmofphere, were extremely difficult and
even laborious, that mathematical exactnefs was not to be
looked for; and that, notwithftanding every poffible precaution was taken, irregularities would occur. Thefe,
however, were not fo numerous as might have been

I　　　　　　　　　　　　　expected,

expected, nor any way ſo great as to render the reſearch
fruitleſs: for a few of that kind being thrown out of the
total number, the mean of the others, which were very
conſiſtent among themſelves, ſerved to prove beyond the
poſſibility of doubt, that the expanſions of common air
did not keep pace with the dilatations of quickſilver.
The manometrical ſpace, anſwering to the 20° of the
thermometer between 52° and 72°, was always found to
be greater than any other 20° of the ſcale. Here it is to
be underſtood, that I do not pretend to have aſcertained
the exact point in that ſpace where the maximum falls:
the mean correſponds to the 62d degree, and yet I am
inclined to think that it is ſomewhat lower, perhaps it
may be about the 57th: from this point, the condenſa-
tions of air downwards, and its expanſions upwards, follow a diminiſhing progreſſion, compared with the con-
denſations and dilatations of quickſilver. The manometrical are equal to the thermometrical ſpaces, in two
points of the ſcale; namely, at or near the freezing temperature on one ſide, and between the 112° and 132d
degrees of the ſcale on the other. At the zero and boiling point they are leſs than the thermometrical ſpaces.
Whether this maximum of expanſion of air, compared
with that of quickſilver, be owing to moiſture, or any

thing.

thing elfe mixed with the former, which is brought into
its greateſt degree of action, about the temperature of
57° of FAHRENHEIT, muſt be left to the inveſtigations
of future experimenters: I only relate things as I found
them after many repetitions, without being able to diſ-
cover any material difference in the reſults, even when
the air was rendered more or leſs denſe by an addition
to, or ſubtraction from, the weight wherewith it was
loaded. The thermometrical, compared with the manometrical ſpaces, will therefore appear as in the following
table.

Spaces of the quickfilver thermometer, FAHRENHEIT's fcale.	Spaces of the manometer, meafured in degrees of FAHRENHEIT.
212°	212°
20	17.6
192	194.4
20	18.2
172	176.2
20	18.8
152	157.4
20	19.4
132	138.0
20	20.0
112	118.0
20	20.8
92	97.2
20	21.6.
72	75.6
20	22.6
52	53.0
20	21.6
32	31.4
20	20.0
12	11.4
0	0

Expe·

Experiments, for determining the actual expansion of common air, in the manometer, affected by the heat of 212°.

For this purpose it became neceſſary to aſcertain, in every manometer, the exact proportion between the capacity of the tube and that of its bulb. This was done, by weighing the quickſilver that filled them reſpectively, in a balance that was ſenſible to a very ſmall fraction of a grain. The contents of the bulb, and that part of the tube between it and zero, expreſſed in grains, was called the air in experiment. The apparent expanſion of that air was meaſured, by the grains that filled the ſeveral ſections of the tube between zero and the boiling point; the ſum being the total expanſion or increaſe of volume, from a heat of 212°. The apparent expanſion, thus found, was again augmented for the dilatation of the tube, on the following principles.

In the firſt part of this paper I have ſhewn, that ſolid glaſs rods dilate much leſs than barometer tubes. The mean between Mr. SMEATON's and my experiments, gives $\frac{14}{1000}$ of an inch for the longitudinal extenſion of every foot of theſe tubes, by 212°. From the rate of going of a clock, for near a year, whoſe pendulum rod is ſolid glaſs, its dilatation ſeems to be one-third part of a ſteel rod,

or

or $\frac{58}{10000}$ on a foot, by 212°. Now, as the manometers re-
femble folid rods much more than they do barometer
tubes, it is probable their dilatation, even allowing for the
greater extenfion of the bulb, would not exceed $\frac{6}{1000}$ths of
an inch on a foot, or $\frac{1}{1000}$th part on every two inches. In
this ratio I have therefore augmented the apparent, to
obtain the true, capacity of each manometer. The equation, amounting to about $\frac{1}{110}$th part of the whole, being
lefs than the common error of fuch complicated obfervations, might in faét have been entirely omitted, without producing any material alteration in the refults.

Having, in this manner, computed the total increafed
volume of any number of equal parts of air (according
to the capacity of the bulb and tube in grains) and very
often likewife the partial expanfions for intermediate
temperatures, expreffed by the contents of the correfponding feétions of the tube, I then found the ratio
anfwering to 1000 equal parts, which, being divided by
the degrees of difference of temperature, gave the mean
rate for the whole fcale, or the particular rate for any
intermediate feétion of it.

The experiments, confidered in this way, are diftributed
into four claffes, whereof the refults are comprehended
in the four following tables. The firft fhews the expanfion
of air, whofe denfity was much greater than that of the

common atmofphere. The fecond, which is divided into two fets, contains thofe on air that fuftained a preffure lefs than the atmofphere. In the third clafs, a very fhort column of quickfilver being employed to confine the air, its denfity differed little from that we commonly breathe in: this clafs is likewife fubdivided into two fets, and it will hereafter be made ufe of to regulate the equation depending on the temperature of the air, in the application of the barometer. The fourth and laft clafs of experiments, were made on air of the common denfity, artificially moiftened by the admiffion, fometimes of fteam, and at others of water, into the bulb; it is accordingly diftinguifhed into two fets.

TABLE I. Refults of experiments on the expanfion of air, whofe mean denfity was equal to $2\frac{1}{2}$ atmofpheres.

N°	Height of the barometer.	Inches of quickfilver confining the air.	Denfity in inches.	Total expanfion of 1000 equal parts of air by 112°.	Mean rate for each degree.
1	29.7	+72.	101.7	451.54	2.12991
2	29.7	+62.6	92.3	423.23	1.99637
3	29.62	+50.84	80.46	412.09	1.94382
4	29.66	+24.88	54.54	439.87	2.07486
5	29.66	+20.05	49.71	443.24	2.09075
		Mean,	75.74	434.00	2.04717

TABLE

TABLE II. Refults of experiments on the expanfion of air of the denfity of five-fix
common atmofphere; and of others on air that was extremely rare, being only pı
about one-fifth of an atmofphere.

	N°	Height of the barom.	Inches of quick-filver.	Denfity in inches.	Total expanfion of 1000 equal parts of air by 212°.	Mean rate for each degree.	Expanfion for intermediate temperatures. From 0 to 32°.	32° to 52°.	52° to 72°	72° to 92°	92° to 132°	13 to 1
Firft fet.	1	29.85	—5.44	24.41	495.455	2.33705	——	——	Not obferved.		——	—
	2	29.76	—3.05	26.71	504.340	2.37896	2.27190	2.41666	2.64060	2.55200	2.46040	2.31
	3	29.79	—0.53	29.26	470.32	2.21849	1.90437	2.48150	2.63150	2.59650	2.15050	2.12
	4	30.09	—5.43	24.66	469.07	2.21259	2.32688	2.53450	2.66250	2.24800	2.25425	2.0
	5	29.93	—9.63	20.30	479.20	2.26038	2.14750	2.49500	2.59850	2.24700	2.25950	2.2
		Mean,		25.17	483.677	2.28140	2.16266	2.48191	2.63327	2.41087	2.28116	2.1

	N°		Difference of temperature.			Tota pan
Firft fet.	6	Experiment in a heated room in Philpot lane, February 25, barometer 30.°03 — 4.82 = 25.21 the denfity of the air, —	on 113½°	from 48½°	to 162°	244
			51½	48½	100	126
			62	100	162	118
Second fet.	7	In Philpot lane; tube with a fmall bore; barometer 30.03—23.2 = 6.77 the denfity. The air had been heated red-hot in the bulb before it was fealed, — — —	on 113½	from 48½	to 162	138
			51	48½	101½	71
			60½	101½	162	66
		The expanfion for 212°, at the mean rate, would be —	—	—	—	25
	8	In Pulteney ftreet, February 28th; with the fame manometer that had been ufed in the fame experiment in Philpot lane, barometer 30.08—23.2 = 6.88 the denfity of the air, — —	on 212	—	—	33
			32 above zero.			4
			20	from 32	to 52	3
			80	52	132	13
			60	132	192	9
			20	192	212	1
	9	In Pulteney ftreet, April 19th; tube with a large bore, barometer 29.8—24.08 = 5.72 the denfity of the air, which had been heated red-hot in the bulb before it was fealed, — —	on 180	from 32	to 212	14
			20	32	52	1
			20	52	72	2
			20	72	92	2
			20	92	112	1
			20	112	132	1
			20	132	152	1
			20	152	172	1
			20	172	192	1

:ratures.		
:°	132° to 172°	172° to 212°
40	2.31850	2.20748
50	2.12000	2.10925
25	2.05325	1.83525
50	2.21375	2.11850
16	2.17637	2.06762

:.	Total ex-panfion.	Mean rate.
	244.604	2.15510
	126.311	2.45264
	118.293	1.90800
	138.75	1.22247
⅓	71.93	1.41039
	66.82	1.10446
	259.164	
	330.487	1.55890
	44.574	1.39294
ι	37 771	1.88855
ι	139.784	1.74730
ι	94.804	1.58007
ι	13.554	0.67770
2	141.504	0 78613
2	17 845	0 89225
ι	25.943	1.29715
2	20.911	1.04550
2	14.937	0.74685
2	14.228	0.71140
2	14.151	0.70755
2	14.150	0.70750
2	12.264	0.61320

20	152	172		1
20	172	192		1
20	192	212		

The Expanſion for 212° at the mean rate would be, — — — — 16

Mean of the three means; denſity 6.46, expanſion for 212°, — — — 2·

2	14.150	0.70750
2	12.264	0.61320
2	7.075	0.35375
	166.660	
	252.104	1.18917

TABLE

TABLE III. Refults of experiments on the expanfion of air of the denfity of the common atmofphere.

N°	Height of the barom.	Inches of quickfilver confining the air.	Denfity in inches.	Total expanfion of 1000 equal parts of air by 112°.	Mean rate for each degree.
1	29.95	+1.57	31.52	483.89	2.28250
2	30.07	+0.70	30.77	482.10	2.27406
3	29.48	+0.42	29.90	480.74	2.26764
4	29.90	+0.83	30.73	485.86	2.29182
5	29.96	+0.96	30.92	489 45	2.30870
6	29.90	+0.65	30 55	476 04	2.24547
7	29.95	+0.65	30.60	487.55	2.29976
8	30.07	+0 53	30 60	482 80	2.27736
9	29 48	+0 52	30.00	489.47	2.30871
Mean			30.62	484.21	2.28401

Firft fet; common air. (rows 1–7)
Second fet; common air heated red-hot (rows 8–9)

The total expanfion 484.21 being divided into parts proportionable to the manometrical fpaces, meafured in degrees of the quickfilver thermometer, as already given; we have the following expanfions for intermediate temperatures, the rates for every 10° below 92° being found by interpolation.

4 X 2   Ther-

Thermo-metrical fpaces.	Manome-trical fpaces.	Total Expanfions for degrees above zero, 1000 parts.	Difference of expanfions, 1000 parts.	Rate for each degree, 1000 parts.
212.	212.	484.210		
			40.199	2.00995
192.	194.4	444.011		
			41.559	2.07795
172.	176.2	402.452		
			42.949	2.14745
152.	157.4	359.503		
			44.310	2.21550
132.	138.	315.193		
			45.680	2.28400
112.	118.	269.513		
			47.507	2.37535
92.	97.2	222.006		
			24.211	2.42110
82.	86.6	197.795		
			25.124	2.51240
72.	75.6	172.671		
			25.581	2.55810
62.	64.4	147.090		
			26.037	2.60370
52.	53.	121.053		
			25.124	2.51240
42.	42.	95.929		
			24.211	2.42110
32.	31.4	71.718		
			23.297	2.32970
22.	21.2	48.421		
			22.383	2.23830
12.	11.4	26.038		
			26.038	2.16983
0.				

Hence 222.006−26.038=195.968=2.4496, or 2.45, is the mean rate of expanfion for the 80° comprehended between 12° and 92° of FAHRENHEIT.

TABLE

TABLE IV. Refults of experiments on the expanfion of air, artificially moiftened, of the manometer.

N°	Height of the barom.	Inches of quickfilver confining the air.	Denfity in inches.	Totalexpanfion of 1000 equal parts of air by 212°.	Mean rate for each degree.	from zero to 32°.	32° to 52°	52° to 72°
1	30.16	+1.6	31.76	——	——	2.059375	2.60700	3.0265c
2	29.97	+2.2	32.17	1409.04	6.64642	2.20250	2.59250	2.9095c
3	30.00	+2.2	32.20	1350.10	6.36840	2.26875	2.59100	3.0490c
4	30.43	+1.92	32.35			2.20875	2.51450	2.7470c
5	30.2	+1.6	31.80	1999.71	9.43259	2.361875	2.51300	2.9640c
6	30.32	+2.37	32.69	2576.16	12.15169	2.16250	2.55350	3.1160c
7	30.2	+1.3	31.50	1135.48	5.35604	2.22594	2.74450	2.9050c
8	30.06	+3.2	33.26	——	——	2.54062	2.63350	2.8085c
9	30 32	+1.6	31.92	1538.31	7.25618	2.02156	2.54250	3.2250c
Mean,			32.18	1668.13	7.86854	2.22799	2.58800	2.9722c
Mean of the fecond, third, and feventh,			31.96	1298.20	6.12362	2.23239	2.64267	2 9545c
Mean of the fifth, fixth, and ninth,			32.14	2038.06	9.61349	2.18198	2.53633	3.1016?

Firft fet: fteam admitted into the bulb before it was fealed. (rows 1–6)

Second fet: a drop of cold water admitted into the bulb before it was fealed. (rows 7–9)

By N° 1. the total expanfion for 192° is 1208.72, when
4. — — 192° 1367.05,
8. — — 112° 358.03,

tened, by the admiſſion of ſteam, and ſometimes water, into the bulb
meter.

Expanſion for intermediate temperatures.							
52° to 72°	72° to 92°.	92° to 112°.	112° to 132°	132° to 152°.	152° to 172°	172° to 192°	192° to 212°
3.02650	3.38050	4.18300	6.48000	8.67750	11.93600	16.85050	——
2.90950	3.67650	5.16700	6.93300	10.17500	10.64200	16.57850	8.25400
3.04900	3.77550	4 36900	7.60500	8 94400	10.42950	11.92200	11.690c0
2.74700	3.25500	3.73700	5.91350	9 18950	11.57550	25.88650	——
2.96400	3.84750	5.76100	7.19450	12.29850	16.69750	19.29500	25.23550
3.11600	3.72300	5.53600	7.83900	12.74100	16 74600	27.84350	45.25000
2.90500	3 47750	5.41900	6.16650	7.98850	8.58950	10.93600	4.98600
2.80850	3.78700	4.60750	——	Tube broken.		——	——
3.22500	3.76500	5 41700	6.79250	9.14350	9.71100	13.75550	19 93270
2.97228	3.63194	4.91072	6.86550	9.89494	12.04087	17.88344	19.22470
2 95450	3.64317	4.98500	6.90150	9.03583	9.88700	13.14550	8.31000
3.10167	3.77850	5.57133	7.27533	11.39500	14.38483	20.29800	30.13940

2, whence the mean rate is 6.29542.
5, — — 7.12005.
3, — — 3.19669.

From

From the experiments of the firſt claſs it appears, that 1000 equal parts of common air, loaded with two atmoſpheres and a half, being affected with a heat of 212°, expands 434 of thoſe parts; that is to ſay, in its dilated ſtate, it occupies a ſpace bearing, to that which it originally filled, the proportion of 1434 to 1000: hence the mean rate of expanſion of air of that extraordinary denſity is 2.04717 for each degree.

From the firſt ſet of the ſecond claſs of experiments it appears, that 1000 equal parts of air, preſſed only with $\frac{4}{5}$ths of an atmoſphere, and ſuffering a heat of 212°, expands nearly 484 of thoſe parts, whereof the mean rate for each degree is 2.28140. The maximum correſponds to that ſection of the ſcale between 52° and 72°; and the rate for the extremes is leſs than the mean.

But in the ſecond ſet of this claſs, when the confined air was rendered ſo extremely rare as to be preſſed with only one-fifth of an atmoſphere, in which caſe there was a neceſſity for heating it red-hot before it was poſſible to make the quickſilver hang in any tube of a moderate length, the expanſion of 1000 equal parts of air is, by the ſeventh and eighth experiments, diminiſhed to about two-thirds of the uſual quantity; and by the ninth, it is conſiderably leſs, amounting only to 141.5 for the 180°

2   com-

comprehended between freezing and boiling, or 0.78613 for each degree. The maximum still correfponds to the fpace between 52° and 72°; and the minimum is con-ftantly at the boiling point.

From thefe three laft experiments it would feem, that the particles of air may be fo far removed from each other, by the diminution of preffure, as to lofe a very great part of their elaftic force; fince, in the ninth ex-periment, the heat of boiling water applied for an hour together, could only make it occupy a fpace which, com-pared with what it filled at freezing, bears the propor-tion of 1141.5 to 1000.

From the third clafs of experiments it appears, that common air, preffed with a fingle atmofphere, whether taken into the manometer in its natural ftate, or heated red-hot therein, has the fame expanfion with air of only five-fixths of that denfity: for 1000 equal parts of this air expanded 484.21 from 212° of heat, whereof the mean rate is 2.28401 for each degree. By comparing this refult with that of the firft clafs, and again with that deduced from the fecond fet of the fecond clafs, it would feem, that the elaftic force of common air is greater than when its denfity is confiderably augmented or dimi-nifhed by an addition to, or fubtraction from, the weight

6   with

with which it is loaded[f]; for, in the firft cafe, it bears
the proportion of 484 to 434; and in the laft, it is (from
the

(f) This difference between the elaftic force of common air, and that which
is artificially rendered more or lefs denfe, by the addition or fubtraction of
weight, particularly the latter, is truly remarkable, and contradicts the expe-
rience of BOYLE, MARRIOTTE, &c. It could not arife from the adhefion of
the quickfilver to the tube, though in the denfe experiments a column of 72
inches was once made ufe of; becaufe the conftant motion given to the manometer before the fpaces were marked, muft either have prevented any irregularity whatever, or made the apparent expanfion fometimes too great, and at
others too little. But the rare experiments ferve to put this matter out of
doubt; for if the adhefion of the quickfilver to the tube had tended to leffen
the apparent expanfion of the air, beneath the truth in one cafe, it muft have
had a direct contrary effect in the other, and augmented it above the truth,
which it evidently doth not.

Thefe experiments on the expanfion of air lefs denfe than the atmofphere,
were extremely difficult and troublefome; and it was not till after feveral
fruitlefs attempts that, with the affiftance of Dr. LIND, an apparatus was prepared for making them with fufficient accuracy. The veffel employed for this
purpofe was made of the brafs tube of a large telefcope, near four inches in
diameter; it was divided into four pieces, which, when fcrewed together, made
a pot of fix feet in height. This was mounted on a platform laid over the
area rails, for the reception of the manometer, which was placed therein, with
the bulb uppermoft, the lower extremity of the tube paffing through a focket at
the bottom of the veffel, and then through a collar compofed of many thick-
neffes of flannel. By means of a brafs plate and three long fcrews, the collar was
made to embrace the tube fo clofely, as to fuffer very little water to pafs: fuch as
did iffue, oozed off along the fides of a paper funnel, bound round the end of
the tube, without entering into the bore. In this pofition, it required fome degree
of force to pufh the manometer up, or draw it down, till the top of the quickfilver confining the air, juft appeared without the collar, fo as to admit the
fpaces to be meafured, from a fixed point marked on the tube. The veffel being
filled with boiling water, was kept to that temperature by means of lamps fufpended

the mean of three experiments) as 484 to 252, when preffed with only one-fifth of an atmofphere.

The

pended around it. Two thermometers were made ufe of; the long one, whofe ball ftood at the bottom; and a fhort one at the top, that defcended no lower than juft to be immerfed in the water. By fome of the firft of thefe experiments, the lamps not being placed directly at the bottom, water was perceived to be a very bad conductor of heat; for it would boil violently at the top, and the fhort thermometer there would mark 212°, while the long one would only mark 185° or 190° at the bottom. By flow degrees the heat would neverthelefs defcend, and in the fpace of fifteen or twenty minutes the whole column would become of the fame uniform temperature. But when the apparatus was adapted for experiments on air denfer than the atmofphere, in which cafe a plate of tin was foldered over the hole at the bottom, that it might be placed on a ftrong fire, the heat was then greateft below, and the long thermometer would mark upwards of 215°, while the fhort one ftood at 209° or 210°. By defifting from blowing the fire, or removing a part of it, the particles of water fuffering the greateft heat would afcend, mix with the reft, and for fome little time make the whole column of an uniform temperature. But the fire being totally removed, the top of the column in cooling was always hotteft; wherefore, in all thefe experiments, whether on denfe or rare air, great care was taken to mix the water thoroughly.

From Mr. DE LUC's book it appears, that M. AMONTONS found the effect of heat on the air confined in his thermometer, which feems to have been the fame fort of inftrument with the manometer, proportionable to the weight with which it was loaded. By this he could not mean that, being of a double denfity, it had twice the expanfion. I apprehend it muft here be underftood, that the fpaces the air occupied, were inverfely as the weights. That being preffed with a double weight, it only filled half the fpace; or with half the weight, a double fpace. This is no doubt nearly, though not accurately, the law that it follows. From thefe experiments it appears, that there is little difference in the actual expanfion or elaftic force of air, preffed with an atmofphere + or — one-third part: yet, when it is rendered extremely rare, its elafticity is wonderfully diminifhed. There feems likewife to be a vifible diminution in its expanfion, when loaded with two atmofpheres and a half. Some of the tubes that I ufed were near nine

feet

The total expanfion 484.21 refulting from the third clafs of experiments, which are very confiftent among themfelves, being divided into parts proportionable to the manometrical fpaces, as meafured in degrees of the quickfilver thermometer, we have the expanfions for intermediate temperatures, expreffed at the bottom of the third table, where, it is to be obferved, the rates for every 10° below 92° were found by interpolation.

Now as barometrical obfervations will probably never be made in a temperature higher than 92° in the fhade, nor in one lower than 12°, if we fubtract 26.038, the expanfion anfwering to 12°, from 222.006, that which correfponds to 92°, we fhall have 195.968 for the 80 intermediate degrees; or 2.45 for the mean rate on each. This equation, compared with Mr. DE LUC's, bears the proportion of 245 to 210, which is a difference of $\frac{35}{10000}$ on every degree, or one-feventh part of the whole: and though this rate will be found hereafter to

feet long. Had it been poffible to have managed them of double or triple that length, fo as to have admitted the air to be preffed with a column of 18 or 20 feet of quickfilver, I am perfuaded the diminution in the expanfion of air of that extraordinary denfity would have been much more perceptible.

Mr. AMONTONS found, that the condenfation of air in his thermometers kept pace with that of fpirit of wine, which we are told follows a decreafing progreffion with refpect to quickfilver: wherefore his experiments agree with thefe, in making the condenfation of air below 57° follow a decreafing progreffion, when compared with that of quickfilver.

exceed that deduced from the operations of the barometer in extreme temperatures; yet they agree exceedingly
well with each other for the mean heat of the air, when
the barometer will come moſt frequently into uſe.

The fourth claſs of experiments are all that now remain to be mentioned. The bare inſpection of TABLE
IV. will ſhew, how greatly ſuperior the elaſtic force of
moiſt is to that of dry air. It is true indeed, that two
kinds of irregularities preſent themſelves among the reſults: firſt, with regard to the total expanſion for 212°;
and ſecondly, as to the greateſt exertion of the elaſtic
force, which ſometimes ſeems to have taken place before
the air has acquired the heat of boiling water. The
firſt is eaſily accounted for: it muſt have ariſen from different proportions of moiſture being admitted into the
ſame quantity of air, which there was no poſſibility of
aſcertaining, the bulbs and their apertures being of very
different dimenſions. With regard to the ſecond irregularity, I am rather inclined to think that it may have
proceeded from error of obſervation, it being difficult to
determine the accurate temperature near boiling; eſpe-
cially when any part of the air roſe above the top of the
veſſel, which was ſometimes the caſe, notwithſtanding its
extraordinary height. Be that as it may, a very uniform
encreaſing progreſſion will be perceived to take place,

from

from the zero of FAHRENHEIT, as far as 152° or 172°; and even to the boiling point, in thoſe which I eſteem the beſt experiments. By adhering to the mean reſult it will appear that air, however moiſt, having that moiſture condenſed or ſeparated from it by cold, its expanſion differs not ſenſibly from that of dry air. Thus the rate for 32° below freezing 2.22799, is nearly the ſame as in dry air; but no ſooner doth the moiſture begin to diſſolve and mix with the air, by the addition of 20° of heat, than the difference is perceptible: for inſtead of 2.46675, the rate for 20° above 32° in dry air, we have 2.588 for that which is moiſt. In the next ſtep of 20°, the rate for dry air is 2.5809; whereas that for moiſt is 2.97. In this manner the progreſſion goes on continually encreaſing, ſo as to give 7.86854 for the mean rate on each degree of the 212°, which is near three times and a half the expanſion of dry air. And laſtly, the rate for the 20° between 192° and 212°, is twice and one-half the mean rate, and about nine times that which correſponds to the zero of the ſcale: but if the compar">on is drawn from the mean of the fifth, ſixth, and ninth experiments, as being probably neareſt the truth, the total expanſion of moiſt, will be more than four times that of dry air; and the rate for the temperature at boiling,

will

will be nearly fifteen times that which correfponds to the zero of FAHRENHEIT.

I am aware it will be alledged, that the proportion of moifture admitted into the manometer in thefe experiments, is greater than what can ever take place in nature; and therefore, in order to be able to judge of the degrees of expanfion the medium fuffers in its more or lefs denfe, and more or lefs moift ftates, that not only air near the furface of the earth, but likewife that found at the top of fome very high mountain, fhould have been made ufe of. I grant all this: but on the other hand it muft be remembered, that thefe experiments are very recently finifhed; that a good hygrometer (if fuch can ever be obtained) a great deal of leifure time, and the vicinity of high mountains, were all neceffary for the carrying of fuch a fcheme into execution.

It is for thefe reafons, and in hopes that other people will, fooner or later, inveftigate this matter ftill farther, not only by experiments made on the expanfion of air, taken at different heights above the level of the fea in middle latitudes, but likewife on that appertaining to the humid and dry regions of the atmofphere towards the equator and poles, that I have been induced to haften the communication of this paper. In the mean time having proved, beyond the poffibility of doubt, that a wonderful

2　　　　　　　　　　　　　　difference

difference doth exift between the elaftic force of dry and moift air, I may be allowed hereafter to reafon by analogy, on the probable effects this will produce, in meafuring heights with the barometer; leaving it to others, much better qualified, to confider how far it will affect aftronomical refractions. In the following fection I fhall therefore give an account of the barometrical obfervations made in ·Britain, and compare them with fome others made in diftant countries.·

S E C T I O N III.

An account of the barometrical obfervations made in Britain, wherein they are compared with fome others of the fame kind made in diftant countries.·

THE revival of the inquiries into that curious and ufeful branch of philofophy, whereby vertical heights are determined to a great degree of exactnefs, by the preffure of the atmofphere alone, we owe to Mr. DE LUC;· who hath undoubtedly removed many of the difficulties that formerly occurred in the application of the barometer, and thereby encouraged others to attempt to overcome, fome part at leaft, of fuch as remain.·

If

If the rule deduced from the obfervations on Saleve had been abfolutely free from exceptions, and if there had not been particular points in the theory concerning which the ingenious author himfelf feems to have entertained doubts, it would probably have been univerfally adopted, without undergoing any very fcrupulous inveftigation; but the obfervations made at Sun-rifing on Saleve, gave refults that were defective, or lefs than the real height. In certain cafes, the equation for high temperatures, remote from the zero of the fcale, appeared to follow a diminifhing, and in others an increafing progreffion. Hence arofe fome caufes of uncertainty, with refpect to the fpecific gravities of quickfilver and air, and the relative expanfion of one compared with the other; efpecially when the atmofphere happened to have more or lefs moifture diffolved in it. It was doubtlefs from confiderations of this fort, that Mr. DE LUC, in his book, fo ftrongly recommends the making of numerous fets of obfervations, in different countries; that, by the united labours of all, this interefting part of natural philofophy, might be brought ftill nearer to perfection.

With this general object in view, I am now to give an account of the principal barometrical obfervations that have been made in Britain, on heights determined geometrically with great care. Thefe heights are claffed in the

the following lift in fix fets, according to the diftricts of the country wherein they are fituated, and nearly in the order of time in which the obfervations were made.

Height in feet.

		Height in feet.
N° I. Heights in and near London.	St. Paul's Church-yard *(g)*, North-fide, and iron gallery over the dome, — — —	281
	Top of Paul's-ftairs, and the faid gallery, —	324
	Top of Scotland-yard wharf, and the dining-room of the Spaniard on Hampftead-heath, —	422
	Great Pulteney-ftreet, and the faid dining-room, —	352
	Pagoda in Kew gardens, — — —	116½
	Gun wharf in Woolwich Warren, and uppermoft ftory of Shooter's-hill inn *(h)*, — — —	444

(g) Mr. BANKS, affifted by other gentlemen, meafured very accurately with a line the height of the ball of St. Paul's above the floor of the church, which was found to agree, exceedingly near, with that taken from the engraved fection of the building. The diftance of the ball from the dining-room of the Spaniard, was found by a bafe meafured on Hampftead-heath; and their relative heights by the angle of depreffion of the ball, taken with the aftronomical quadrant from the faid dining-room. The heights of Paul's ftairs and Scotland-yard wharf, with refpect to each other, were found by meafuring from them feverally to the furface of high water in the Thames. And the elative heights of the church-yard and floor of the church with refpect to the ftairs, and of Pulteney-ftreet with regard to the wharf, were obtained by levelling to them refpectively.

(h) The height of Shooter's-hill inn, above Woolwich, was found by a bafe meafured in the meadows from the Warren eaftward. Lord MULGRAVE, Mr. BANKS, and Dr. SOLANDER, affifted in the geometrical operations; as did Dr. BLAGDEN, Meff. DE LUC and LLOYD, in the barometrical obfervations.

N°

		Feet.
N° 2. near Tay-bridge in Perthshire.	Station at the East-gate of Castle Menzie's gardens near the village of Weem, and top of Weem Craig,	700¼
	The said station, and top of Bolfracks Cairn, —	1076½
	The said station, and top of Dull Craig, —	1244½
	The said station, and top of Knock Farle, —	1364½
	The said station, and that at the rivulet of Glenmore, below the South observatory on Schihallien, —	1279¼
	The said station, and South observatory, — —	2098
	The said station, and Western summit of Schihallien,	3281
	Station at the rivulet of Glenmore, and the South observatory, — — — —	818¾
N° 3. near Lanark.	Level of the Clyde at Lanark-bridge, and station in the garden at Lanark, — —	362½
	Ditto level, and top of Stonebyre-hill, — —	654
	Robinhood's well, before Carmichael-house, and top of Tinto, four feet below the summit of the Cairn,	1642½
	Ditto well, and West end of Carmichael-hill, —	451½
N° 4. near Edinburgh.	Leith pier-head, and top of the Calton-hill, —	344
	Leith pier, and summit of Arthur's Seat, — —	803
	Leith pier, and Kirk-yetton Cairn, on the East-end of the Pentland hills, — — —	1544
	Calton hill, and ditto Cairn, — — —	1200
	Level of Hawk-hill study, and top of Arthur Seat,	702½
	Hawk-hill observatory, and bottom of the little rock on Arthur Seat, 7¼ feet below the summit, —	684
	Hawk-hill garden-door, and ditto little rock, —	730¼
N° 5. near Linhouse.	Linhouse, and East Cairn-hill, 5 feet below the summit,	1176½
	Ditto, 18 feet below the top, — —	1165½
	Linhouse, and West Cairn-hill, 11 feet below the top,	1178½
	Ditto, and Corstown-hill, 4 feet below the top, —	386½
	Corstown-hill, and West Cairn-hill, — —	792
	Ditto, and East Cairn-hill, — — —	776½
N° 6. near Carnarvon in North Wales.	Carnarvon Quay, and Snowdon Peak, — —	3555
	Ditto, and summit of Moel Eilio, — —	2371

To enter into a minute detail of the geometrical operations, whereby the whole of these vertical heights were

were determined, would be extremely tedious and unin-
terefting. That fome idea may however be formed of
the degree of accuracy with which they were afcertained,
it will be fufficient to obferve, that the requifite angles
were taken with an aftronomical quadrant of a foot ra-
dius, made by Mr. sisson, and curioufly adapted for the
meafurement of horizontal or bafe angles; which, as well
as thofe of the vertical kind, might always be determined
thereby to within ten feconds of the truth. The bafes
were meafured with care; and, in order to afcertain the
diftances, the three angles of each triangle were, as
often as poffible, actually obferved with the quadrant.
That the variation of the line of collimation of the inftrument, which was found to alter in carrying, might
occafion no error, one or more of the angles of elevation,
at each ftation, were taken on the arc of excefs, as well as
on the quadrantal arc. In all cafes, the ufual [i] allow-
ances were made for curvature and refraction: and for
the correction of the laft, fometimes the angles of de-

(i) If the fquare of the diftance be divided by the diameter of the earth, the
quotient will give the curvature of the globe on that diftance, or the excefs of
the apparent above the true level: and, by Mr. MASKELYNE's rule, the fquare of
the diftance being divided by the diameter of the earth, augmented by one-
fourth part, we have the allowance for curvature and refraction; which laft is
fuppofed to raife the object, by an angle equal to that of a great circle fubtended
by one-tenth part of the diftance.

VOL. LXVII. 4 Z preffion

preffion as well as of elevation were taken. When time would permit, the geometrical operations were repeated at the firft ftations; or the angles of elevation were obferved from fome new point connected with the firft, and whofe relative height, with refpect to the others, was known. Small altitudes were occafionally determined by levelling from one ftation to the other.

To prove that the vertical heights, affigned to the places in the preceding lift, are exceedingly near the truth, I need only mention the following inftances, by way of confirmation. In 1771, with the affiftance of Dr. LIND and his friend Mr. HOY, I meafured a bafe from the obfervatory of Hawk-hill weftward, whereby the height of the fummit of Arthur's Seat, above the telefcope of the Hawk-hill quadrant, in its horizontal pofition, was found to be 685.66 feet. In 1775, thefe gentlemen levelled, three feveral times, from the fummit downwards to the faid telefcope; and found the vertical diftance to be, by the firft operation, 686.47 feet; by the fecond, 684.43; and by the third, 685.25. This laft, which, from the great care that was ufed, they confidered as the beft, differs only three-tenths of a foot from the geometrical refult. They afterwards continued the operation of levelling from Hawk-hill to the pier of Leith, and having repeated it twice, with a difference of only two inches

between

between the refults, they found the mean defcent to be
117.38 feet: hence Arthur's Seat is above Leith pier, by
the mode of levelling, 802.66; and by the mixed method 803 feet.

In 1774, when the aftronomer-royal was carrying on
the Society's experiments for afcertaining the attraction
of Schihallien, I found, from my own geometrical operations, depending on a bafe meafured in the plain near
Taybridge, the Weftern fummit of the mountain to be
1183 feet above the South obfervatory.

Of this height, the effect of curvature and refraction
amounted to 28.86 feet, on the diftance of Bolfracks
Cairn from the obfervatory; and to 38 feet, on the diftance of the faid Cairn from the fummit of Schihallien.
The refult of thefe operations I communicated to Mr.
MASKELYNE, before his trigonometrical operations were
begun. From the data which he hath fince been fo
obliging as to furnifh me with, depending on the bafe in
Glenmore at the bottom of Schihallien, and the angles
of elevation taken from the Southern extremity of that
bafe, the Weftern fummit of the mountain is 1186.6
feet above the South obfervatory. But if the triangle
that ferved to connect the ftation of the barometer in
that valley with the others, and the angles of elevation
taken from the fame ftation are made ufe of, the dif-

4 Z 2   ference

ference of height will be 1183.33 feet. Laſtly, Mr.
MASKELYNE's reſult, from the triangles on the North-ſide
of the mountain, makes it 1180. The mean of theſe
three 1183.31 feet, is the height of the Weſtern ſummit
of the mountain above the South obſervatory, which only
exceeds my height by one-third of a foot. Here it is to
be obſerved, that from the vicinity of theſe triangles to
the mountain, and the ſhortneſs of the ſides, the greateſt
curvature amounted only to 16 or 17 inches, which con-
ſequently made the effect of refraction next to nothing.
This near agreement between the reſults ſeems therefore to prove, that the mode of computation for curvature and refraction, made uſe of in the Taybridge obſervations, is juſt.

By the firſt angles of elevation, taken from the ſtation
of the barometer in Glenmore, the Weſtern ſummit of
Schihallien is 2001.88 feet above it; from which, if we
deduct 1183.33, there remains 815.55 for the height of
the South obſervatory above the ſaid ſtation: but if the
laſt angles of elevation at the ſtation of the barometer
are made uſe of, the height between it and the obſervatory comes out 818.97, whereof the mean is 818.76
feet. Though theſe inſtances are of themſelves ſuffi-
cient to prove, that the geometrical heights may be ſafely
depended upon; yet, as an example of the method that

was

was always made ufe of, I fhall annex to this paper, a plan of the triangles and detail of the operations for obtaining the height of Snowdon; becaufe that mountain, at the fame time that it is the higheft I have meafured, is, from its fituation, more likely to be vifited, and to have experiments repeated upon it, than the remote hills of the North. I now proceed to give fome account of the barometrical obfervations.

The heights in and near London being fo very inconfiderable, it was eafily forefeen, that nothing conclufive could be drawn from obfervations made on them alone. It was, however, natural enough to try, even on thefe, whether the rule we had been furnifhed with would anfwer? A fmall height of 41 feet 4 inches, which, without inconveniency, could be recurred to at all times of the day, and all feafons of the year, was the firft that was made ufe of. St. Paul's, Hampftead, Kew pagoda, and Shooter's-hill, were the next. The mean refults of many obfervations on the three firft, and of feveral on Shooter's-hill, were found to be defective. In general the coldeft obfervations, made in the morning and evening, when the temperatures at the two ftations differed leaft from each other, anfwered beft. In the hotteft part of the day, when that difference was the greateft, the refults were moft defective.

Some

Some months fpent in Scotland in the fummer of
1774, afforded opportunities of making barometrical obfervations on hills of various heights, from three or four
hundred, to upwards of three thoufand feet, as hath been
exhibited in the preceding lift.   That feafon was remarkably cold and wet; wherefore, in thefe obfervations,
the mean temperature of the air in the fhade was commonly about 55°.   The hotteft never exceeded 63° in
the plain; and the coldeft, namely thofe on the higheft
mountains, were generally from 43° to 48°.

From the defect found in the refults of thefe obfervations, which, with refpect to temperature, correfpond to
the mean and hotteft of thofe made at Sun-rifing on Saleve, and without any exception whatever, I could eafily
difcover, either that a much greater equation than what
the rule directed, muft be applied for each degree of heat
above the zero of the fcale; or, that the zero itfelf would
fall confiderably lower than 39°74, where Mr. DE LUC's
formula, adapted to Englifh meafures, hath fixed it.
This firft ftep towards a correction of the rule, naturally
pointed out the fecond thing to be aimed at, namely, the
obtaining of a fufficient number of cold obfervations,
near the zero, and as far as poffible below it, that the
equation might difappear entirely, and even come to be
applied with the contrary fign.   Of this kind the winter

I   feafons

feafons of 1774 and 1775 afforded a few on the fmall heights in and near the metropolis; but the beft I have been furnifhed with are thofe which Dr. LIND, affifted by Mr. HOY, was fo obliging as to make on Arthur's Seat near Edinburgh; and thofe which Captain CALDERWOOD has fince favoured me with on the Cairn-hills, being a part of the Pentland range to the South-weft of that city.

By comparing thefe fets of obfervations together, it appeared from all of them, that when the air was at or near the freezing temperature, the logarithmic differences gave the real height, in Englifh fathoms and thoufandth parts, without any equation; and when confiderably below that point, the equation was to be fubtracted, or applied with the fign – inftead of +. It was farther perceived, that the fame general conclufion might be drawn from the coldeft, not only of the Sun-rifing, but even of the ordinary obfervations on Saleve; fome reduction of the temperature being in certain cafes made, on account of the expofure of the thermometer to the Sun's rays: hence I was led to fuppofe, that the morning obfervations, inftead of being made exceptions from the rule, were thofe, which, it might be prefumed, would form the beft bafis for deducing the equation depending on the heat of the air; becaufe the mean temperature of the column was then found to differ leaft from that of it's

extre-

extremities; whereas in the hotteſt time of the day, that difference was generally the greateſt.

Having been enabled, by means of the cold obſervations, to form ſome judgement whereabout the zero of the ſcale would fall, below which the equation was negative, and above it affirmative; it followed of courſe, that the next principal thing to be ſought for, was the maximum of equation, or that correſponding to the higheſt temperatures the climate of our iſland would afford. It was partly with the view of obtaining theſe that I went, in July 1775, to Snowdon in North Wales. On this expedition Captain CALDERWOOD was ſo obliging as to accompany me, and lend me his aſſiſtance in the operations for determining the geometrical height of that remarkable mountain. At that particular period, the weather proved unfavourable for obtaining hot barometrical obſervations; but, in other reſpects, they were very ſatisfactory, as being in general conſiſtent among themſelves, and agreeing ſufficiently near with thoſe of the preceding year in Scotland; at the ſame time that they were made on a height, as formerly mentioned, greater than any other hitherto meaſured, with equal care, in Britain. During the ſummer of 1776, Dr. LIND obtained ſome more hot obſervations on Arthur's Seat; and in the beginning of the following winter, Captain CALDERWOOD

5   made

made others of the cold kind, on the Cairn-hills in his neighbourhood. From the combination of the whole of thefe obfervations taken together, and a comparifon of them with Mr. DE LUC's, as far as they are fimilar, I mean to fhew the agreement or otherwife, between the equation for the heat of the air, as deduced from the barometer and manometer; but fince the Britifh obfervations, in certain cafes, differ confiderably in their circumftances from thofe on Saleve, it is neceffary, in the firft place, to point out wherein this difference chiefly confifts.

In the obfervations in Britain, the barometers and detached thermometers have been, almoft conftantly, placed in the open air in the fhade, and fuffered to remain there generally half an hour, and fometimes a great deal longer, before the correfponding obfervations were begun, that the quickfilver might have time to take the temperatures of the fituations refpectively. They were then obferved four times, ufually at intervals of ten minutes, the mean of the four being that which is calculated, and called a fingle obfervation. If the time did not admit of fo long an interval, the fame number of obfervations were taken at diftances of five minutes from each other. In either cafe, the extremes never differed above a few thoufandth parts of an inch from the mean, fo as to

render the computations of them feparately wholly unneceffary.

Except in very fmall heights, and chiefly in London,
where it was impoffible to fcreen the upper barometer fo
effectually from the Sun during the time of obfervation
as that below, which generally ftood in the fhade of fome
building, the temperature of the quickfilver in the fuperior [k] hath been colder than that in the inferior barometer. The difference was commonly found to be two
or three degrees; fometimes it would amount to fix
or feven; rarely, in heights that were confiderable, to
nine or ten; and in one inftance only to thirteen, where
the vertical diftance of the inftruments was great.

Whether in the plain or on the tops of the higheft
mountains, the detached thermometers, indicating the
temperature of the air, have generally ftood fomething

(k) I have fometimes found, particularly in frofty weather, that a thermometer placed on the pavement of the North-fide of St. Paul's Church-yard,
clofe to the wall of the building, would ftand two degrees lower than that which
was expofed on the North-fide of the iron gallery over the dome. The firft, no
doubt, felt the cold produced by the evaporation from the ftones, while that
above might be affected by the afcending fmoke. But the moft remarkable
inftance of this kind occurs in one of Dr. LIND's obfervations, on the breaking
up of the hard froft January 31, 1776: at Hawk-hill, at 10ʰ 45″ A.M. the
temperature of the open air was 14°, while that at the fummit of Arthur's Seat
was 20°. The froft that remained in the ground kept the air extremely cold
below, though it had already felt the effects of the thaw at the top of the
mountain.

                                                        lower

lower than thofe attached to their refpective barometers, until they had remained a confiderable time in the fame fituation, equally fhaded from the Sun, when they always agreed: whence it followed, that in thefe obfervations, the mean temperature of the air, and equation depending upon it, might always have been determined very near the truth, from the temperature of the quickfilver in the tubes, as fhewn by the attached thermometers, without ever confulting the detached ones. Let us now fee what were the circumftances attending the obfervations on Saleve.

Mr. DE LUC's lowermoft barometer ftood in the ground-ftory of a houfe near Geneva, where it remained unaltered during the whole of his experiments; while the detached thermometer, indicating the temperature of the air, was expofed on a fmall eminence, at a little diftance, directly to the Sun's rays: hence we find that, in the obfervations of high temperatures, the bottom of the column of air is often 12° or 15°, and in one cafe 18°, hotter than the quickfilver in the tube. And even in the loweft temperatures, the bottom of the column of air is generally hotter than the quickfilver within doors, contrary to common experience in this country: for in England, in winter, the exterior air in the fhade is always colder than the interior air. This circumftance gives

reafon

reafon to apprehend, that the thermometer fuffered not only direct but reflected heat.

The fuperior barometer was fhaded with a parafol from the Sun, while its correfponding detached thermometer was expofed to his rays: wherefore, in the obfervations of high temperatures, the top of the column of air is ufually four or five degrees hotter than the quickfilver in the barometer ftanding in the fame air; and the mean heat of the column often exceeds very confiderably the mean heat of the quickfilver in the tubes.

In many of the coldeft of Mr. DE LUC's obfervations, as well as in thofe of mean temperatures of about 50° or 55°, the fuperior barometer is often the hotteft of the two, even when the furrounding atmofphere at the top is colder than at the bottom. This circumftance is eafily accounted for: wood is known to be a bad conductor of heat, to receive it flowly, and retain it long: that barometer, which was moved about from place to place upon the mountain, with a very fhort interval between the obfervations (as is fufficiently evident from the great number of ftations it paffed through in a limited time) muft have acquired and retained heat fuperior to that of the atmofphere, and communicated it to the tube with which it was in contact. Some difference would no doubt arife from this caufe, if the temperatures of the

quickfilver

quickfilver in the tube and attached thermometer did not keep exactly pace with each other.

The laft point to be mentioned is ftill more remarkable than the reft; it is briefly this: in the obfervations at Sun-rifing on Saleve, though the fuperior quickfilver is the coldeft; yet the top of the column of air is commonly five or fix, and fometimes eight or nine degrees, warmer than the bottom.

Having thus fhewn the fteps that were taken, for obtaining the coldeft and hotteft barometrical obfervations that the climate of this ifland would afford, the mode of obferving, and wherein the circumftances attending them differed from thofe on Saleve, I fhall now point out the general refult. In order to avoid repetitions as much as poffible, it is neceffary, once for all, to remark, that the computations of the Britifh obfervations, by the rule hereafter to be given, are fubdivided into their refpective claffes. Each table contains 15 columns, which their titles fufficiently explain, that the principles from which the rule was deduced, the refult and error, might all appear in one view. The laft column towards the right-hand fhews the ratio of the weight of quickfilver to air, the columns of the firft in the barometers being feverally reduced to the mean temperature of the laft.

2   By

By comparing the tables it will be found, that the obfervations for extreme temperatures belong to the Edinburgh clafs of obfervations (N° 4.) it being thought beft, in this cafe, to omit the few hot ones obtained on the inconfiderable heights near London: the mean of the coldeft, anfwering to the temperature of 21°.75, make the logarithmic excefs $\frac{29}{1000}$; and the mean of the hotteft, correfponding to the temperature of 69°.6, give a defeft of $\frac{81}{1000}$. Now the fum of the two equations $\frac{110}{1000}$, being divided by the difference of temperature 47°.85, we have nearly 2.3 for the mean rate of the equation on each degree, which is lefs than that refulting from the operations of the manometer. Again, from the mean of the very beft obfervations, as being made on the greateft heights, when the temperature of the air is 52°, it appears, that the defeft is from $\frac{49\ to\ 50}{1000}$, or 2.5 for each degree nearly, which agrees perfectly well with the manometrical expanfion. In this cafe, the ratio of the weight of quickfilver to air is as 11377 to 1; greater very confiderably than 11332 to 1, affigned to them by Mr. DE LUC, when the temperature is 69°.32, anfwering to the zero of his fcale, without any allowance for the diminution of preffure on his columns, which fhould have rendered air ftill comparatively lighter. From the Britifh obfervations, made on the moft confiderable heights,

it

it appears, that when the temperature of the air is 28°.2, the ratio of its weight, with refpeЄt to that of quickfilver, is as 1 to ⸱0552: hence the increafe of the weight of air, on every degree of difference of temperature between 28°.2 and 52°.5, amounts to 34.4; and hence we have 52°.5−4°.2=48°.3 for the temperature of the air in Britain, when its weight would be $\frac{1}{11332}$ of that of quickfilver; and confequently agree with Mr. DE LUC's, though the heat would differ from his 21°. It will no doubt be remarked, that the equation for the air, refulting from the operations of the barometer, falls fhort of that given by the manometer. Part of the difference, I apprehend, may arife from the fmall number of barometrical obfervations obtained in extreme temperatures. I fhall, neverthelefs, adduce reafons hereafter for fuppofing, that it really fhould diminifh, becaufe of the drier and lefs elaftic ftate of the fuperior air, compared with that taken into the manometer at the earth's furface. In the mean time, fince both inftruments agree in the equation for 52°, which is a heat that the barometer will very frequently be ufed in, it feems beft to adhere to the mean manometrical refult 2.45, in fixing the zero of the fcale, which is obtained in the following manner.

Divide the excefs or defeЄt, exprefled in 1000th parts of the logarithmic refult, by 2.45, the mean expanfion of

air for each degree of the thermometer; the quotient
will give the number of degrees, in the firſt caſe, to be
added to, and in the laſt ſubtracted from, the temperature of the air in the obſervation; the ſum or difference
anſwers to the zero of the ſcale, or that temperature
when the logarithmic reſult gives the real height in
Engliſh fathoms and 1000th parts.

According to this mode of computation, we have,
from the aggregate of the ſeveral claſſes of Britiſh obſervations, the place of the zero as follows:

Zero.

By the 1ſt claſs of obſervations in and near London, between the temperatures of — —	}	25.5 and 71.2	at 32.2
2d, near Taybridge, — —		46.1 — 62.9	– 31.1
3d, near Lanark, — —		44. — 62.	– 32.8
4th, near Edinburgh, —. —		17. — 70.7	– 31.3
5th, near Linhouſe, — —		26.1 — 46.5	– 29.9
6th, near Carnarvon, — —		49.1 — 62.3	– 32 9
Mean place of the zero at — —			31.7

The number 31°.7 differing ſo very little from 32°,
we may hereafter conſider that remarkable point of FAH-
RENHEIT's thermometer, as the zero of the ſcale depending on the temperature of the air; and hence is deduced
the ſecond part, of the rule for meaſuring heights with
the barometer. When the mean temperature of the column of air to be meaſured is at 32° of FAHRENHEIT, the
difference

difference of the common logarithms of the equated
heights of quickfilver in the inferior and fuperior barometers, expreffed in 1000th parts of an inch, gives the real
height in fathoms and 100th parts, the three figures
towards the right-hand being decimals, and the reft integers; which, being multiplied by fix, gives the refult
in feet.

Let us next confider, in a general way, how far this
will correfpond, or otherwife, with Mr. DE LUC's obfervations in extreme temperatures.

I have already had occafion to remark, that when the
temperature of the air was at 69°.32, as indicated by
thermometers expofed to the Sun's rays, Mr. DE LUC
found that the differences of the common logarithms of
the heights of the barometers at the two ftations, gave the
altitude between them, in French toifes and 1000th
parts: in which cafe the fpecific gravity of quickfilver to
air was as 11232 to 1. When his formula is adapted to
Englifh meafures, the zero of the fcale neceffarily de-
fcends to 39.74, where the Englifh fathom bears the
fame proportion to the modulus of the common logarithms, as, in the former cafe, the French toife did to that
modulus, the equation for the intermediate temperature
being now applied with the contrary fign. As it hath
been fhewn, that the Britifh obfervations differ in their

circumftances from thofe on Saleve, and require a greater
equation, it is unneceffary to enter into any minute comparifon of the two fets: neverthelefs, that fome idea may
be formed of the caufe, of part at leaft, of the difference
that takes place between them, I have collected into one
view, the computations of fuch as were made in extreme
temperatures; namely, the coldeft of thofe at Sun-rifing
(though the whole of that clafs were confidered as exceptions from the rule); the coldeft and hotteft of the
ordinary obfervations; alfo thofe on the Dole[1], at Genoa,
and at Turin, whereby the heights of the lake of Geneva and of Turin, above the fea at Genoa, were obtained. In the table it will be obferved, that there is a
column for the reduced temperature of the air, on account of the expofition of the thermometer to the Sun's

(1) Having recomputed the whole of Mr. DE LUC's obfervations on Saleve,
and claffed them according to the months in which they were made, I intended,
at one time, to have given a general table, comprehending the mean refults of
all of them: however, this is now become unneceffary, fince a very refpectable
and ingenious member of this Society hath had opportunities of making many
curious and interefting obfervations on thofe very heights, which cannot fail of
being perfectly fatisfactory; and who, at my requeft, was fo obliging as to
determine the height of the Dole geometrically. On this mountain Mr. DE LUC
had made barometrical obfervations, whofe refults differing confiderably from
the altitude, 4182 feet above the lake of Geneva, as taken by Mr. FATIO DE
DUILLIER, made me fufpect there was an error. In fact it appears, from the
laft meafurement, that the fummit of the Dole is 4293 feet above the furface of
the lake, which gives for the vertical diftance of Mr. DE LUC's barometer
4210 feet.

rays:

rays: I apprehend that I have not exceeded, but rather fallen fhort, in the reduction, to what would have been indicated by thermometers in the fhade, perfectly free from direct and reflected heat, and with fufficient time allowed between the obfervations. Be this as it may, it is of no importance, as no other conclufion is drawn from thefe obfervations, than that of fhewing what, in my apprehenfion, might probably have been the cafe, if another mode had been adopted.

From the table it appears, that when the temperature of the air is at $29^{\circ}.5$, the logarithmic excefs is $\frac{9}{1000}$; and at $75^{\circ}.5$ reduced temperature, the defect is $\frac{96}{1000}$. The fum of the two equations $\frac{105}{1000}$ being divided by the difference of temperature 46°, we have, as in the Britifh obfervations, nearly 2.3 for each degree, which is greater than that applied by Mr. DE LUC's rule, in the proportion of 23 to 21. That too fmall an equation hath been made ufe of in thefe hotteft obfervations, fuppofing the original zero and temperature to remain, is fufficiently evident: for $\frac{96}{1000}$ being divided by 42° the difference of temperature, we have, as before, 2.3 very nearly for the equation anfwering to each degree.

Farther, if we confider the ratio of the weight of quickfilver to air, actually refulting from the obfervations themfelves, the fame kind of error (for I cannot fee

it

it in any other light) ftill exifts. Thus, in the coldeft of the morning, as well as in the ordinary obfervations, when the temperature is at or near freezing, the mean ratio of the weight of quickfilver to air, is about 10850 to 1. When the obferved and reduced temperatures are re-fpectively 41° and 35°, the ratio between them is that of 11295 to 1, anfwering nearly to what hath been af-figned to them when the heat is 69°.32. Again, in the hotteft obfervations of the 14th and 15th of July 1759, and 20th of July 1760, on the higheft, and confequently the beft ftations, when the obferved and reduced temperatures are refpectively 81°.7 and 75°.6, quickfilver is to air as 12650 to 1. Now if we reduce this number 12650 by a proportionable part, for the degrees of difference between Mr. DE LUC's zero 69°.32, and the obferved and reduced temperatures re-fpectively, we fhall have, in the firft cafe, 12200; and in the laft, 12390 to 1, for the ratio of the weight of quickfilver to air; either of which exceeds very confi-derably 11232, which hath been affigned to them.

With regard to the obfervations on the Dole, the de-fect is $\frac{814}{1000}$, anfwering to the obferved temperature of 66°.6, and which is only reduced to 65°.2. On this great height, the ratio of the weight of quickfilver to air[a] is that of 12595 to 1. Mr. DE LA CAILLE's obfer-vations at the Cape of Good Hope, annexed to the table containing Mr. DE LUC's, give a defeét of $\frac{18.7}{1000}$, when the temperature feems to have been about 58°, in which cafe quickfilver was 11687 times heavier than air.

Now if, from the aggregate of thefe obfervations, the fame method be adopted, as was made ufe of in the Bri-tifh, for finding the zero of the fcale, we fhall have it as follows:

By Mr. DE LUC's equation for the air and obferved temperature.				By the manometrical equation and reduced temperature.			
	°	°	°		°	°	°
Coldeft of the morn-ing obfervations, from	25.2 to	30.5 at	33.7	from	25.2 to	30.5 at	33.12
Coldeft of the ordi-nary obfervations,	27.1 –	41.9 –	38.7	—	26. –	35. –	32.97
Hotteft of the ordi-nary obfervations,	76. –	84.5 –	36.2	—	73.5 –	77. –	36.32
On the Dole, —	59.2 –	71.5 –	27.6	—	58. –	70. –	32.
Light houfe of Genoa,	75. –	79. –	26.	—	75. –	79. –	33.40
DE LA CAILLE's, —		58. –	30.	— —		58. –	33.35
Zero at — —			32.03	Zero at			33.52

(m) It will even be found, though the calculations are not inferted in the table, that the hotteft of Mr. DE LUC's morning obfervations, June 8th, 1758, at the 15th ftation, anfwering to the mean temperature of 57°.5, and which I confider as the beft, becaufe no reduétion is neceffary for the expofition of the thermometer, agree with the manometrical experiments, in requiring a greater equation than is wanted in extreme temperatures: for in this cafe, the mean of two obfervations gives a defeét of $\frac{65.5}{1000}$ for 25°.5 above freezing, which is 2.57 for each degree; the ratio of the weight of quickfilver to air being that of 12196 to 1.

From

From the mean of thefe obfervations, though the refults are irregular among themfelves, it appears fufficiently evident, that if the morning obfervations on Saleve had been retained, inftead of being made exceptions
from the rule, the zero of the fcale would have defcended about 8°; *viz.* from 69°.3 to 61°.4 of FAHREN-
HEIT, fuppofing always the equation 2.1 for each degree
of temperature, and the French toife, as the ftandard
meafure, to have been adhered to: for the French toife
bears to the Englifh fathom, the proportion of 106575
to 100000; wherefore $\frac{6575}{106575} = \frac{61.69}{2.1} = 29°.4 + 32° = 69°.4$,
denotes the relative pofitions of the two zeros, the inter-
mediate equation $\frac{61.69}{1000}$ being to be fubtracted, when the
toife is made ufe of. But it hath been fhewn, that the
mean expanfion of air is really greater, for fuch temperatures at leaft as the barometer can be applied in, than
what Mr. DE LUC fuppofed it, in the proportion of 245 to
210; whence it follows, that $\frac{61.69}{245} = 25°.18 + 32° = 57°.18$,
will denote the relative pofitions of the two zeros: which,
inftead of almoft 30°, are only diftant from each other a
little more than 25°.

From what hath been faid it is eafy to fee, that in calculating heights according to Mr. DE LUC's rule, when
the temperature of the air is below his zero, which we
                              may

may take at 40°, the Englifh meafure being ufed, the
common error in the refult will be equal to the fum of
the two equations, $2.1 + 2.45 = 4.55$ for each degree;
which amounts to $\frac{164}{1000}$ parts for the 8° that the zero is too
high. Above 40°, the former error $\frac{164}{1000}$ will be augmented
by the difference of the equations for each degree that
the temperature is above his zero, *viz.* $2.45 - 2.1 = \frac{035}{1000}$.
In either cafe it is to be obferved, that the progreffive
rate of equation for the heat of the quickfilver is not
here taken into the account; becaufe it will not produce
any material difference, unlefs one barometer is much
hotter than the other, at the fame time that their vertical diftance is very great. Thus the 32d degree of FAH-
RENHEIT, or freezing temperature, which is fundamental in all thermometers, happens, fomewhat remarkably,
to be the zero of the fcale, when the Englifh fathom
bears fuch proportion to the modulus of the common logarithms, as that their difference, in computing heights
by the barometer, brings out the refult in fathoms. No
other proportion of a meafure will do it: for if we fuppofe twenty-four of different lengths, between ours
and the French toife, each furpaffing the other by $\frac{163}{100000}$
of that toife, the zero of the fcale, in computing heights
by thefe meafures refpectively, will afcend a fingle de-

gree

gree on each; and the French toife being the 25th, will
have its zero nearly at the 57th degree: about which
temperature the expanfion of air appears, from the experiments, to be at its maximum. From that point,
therefore, the equation will diminifh both ways, though
by a quicker progreffion for condenfation, than it doth
for dilatation.

Having thus compared, in a general way, the refults
of the Britifh obfervations with thofe of Mr. DE LUC,
pointed out what feem to be the chief caufes of the conftant defect found in his rule, and thereby obtained, it is
hoped, fome corrections tending to improve the theory
of the barometer, when applied to the meafurement of
heights in middle latitudes; it remains to fhew the principles, whereon the table for the equation of the air hath
been conftructed. Previoufly however to this, it may be
proper to compare, with as much brevity as poffible,
thefe obfervations, with others that have been made towards the Pole and at the Equator: from which it will
appear probable, that the rule which anfwers in middle
latitudes, will not in the frigid and torrid zones.

In 1773, Captain PHIPPS, now Lord MULGRAVE, com-
manding two of his Majefty's fhips then fent on difcove-
ries towards the North Pole, meafured geometrically, with
great care, the height of a mountain in Hakluyt's Ifland

near

near Spitzbergen, and found it to be 1503 feet above the level of the fea. On the morning of the 18th of Auguft, the following obfervations, at the fea-fhore and top of the mountain, were made with a fingle barometer, wherein the quickfilver had not been boiled.

At 6 h. A.M. Barometer at the fhore, — — 30.040 therm. 50°
7 h. 45' A.M. Ditto at the top of the mountain, 28.266 —— 42
8 h. 45' A.M. Ditto at the top of ditto, — 28.258 —— 42
11 h. 45' A.M. Ditto at the fea-fhore, — 30.032 —— 44

Whence we have the following computations, equated for the times correfponding to the two obfervations at the top.

$$7 \text{ h.} 45' \text{ A.M.} \begin{cases} 30.038 & 46°-046=29.992 \\ 28.266 & 42-031=28.235 \end{cases} = 1573 \begin{cases} +70. \\ =44.5 \end{cases} \begin{cases} 43° \\ 41 \end{cases} 42°$$

$$8 \text{ h.} 45' \text{ A.M.} \begin{cases} 30.036 & 45°-043=29.993 \\ 28.258 & 42-031=28.227 \end{cases} = 1581.3 \begin{cases} +78.3 \\ =49.5 \end{cases} \begin{cases} 43 \\ 41 \end{cases} 42$$

Mean $1577.1 = +\dfrac{47}{1000}$

Quickfilver to air as 10224 to 1.

From thefe obfervations it appears that, inftead of the ufual equation $\frac{45}{1000}$, to be added to the logarithmic refult, in order to obtain the true height in Britain, when the temperature is 42°, there is an excefs of $\frac{47}{1000}$: and, inftead of the ufual ratio of the weight of quickfilver to columns of air, of equal altitude and temperature in Britain, namely about 11200, we have that of 10224 to 1.

Thus

Thus air at Spitzbergen feems to be fpecifically hea‑
vier, than that affected with the fame heat and preffure
in the middle latitudes: whence it follows that, inftead
of 32° which is found to be the zero of the fcale about
the middle of the temperate zone, we fhall have $\frac{47}{145}$ =
19°.2 + 42° = 61°.2 for the zero at Spitzbergen, within
10° of the North Pole.

It is much to be regretted, that the French academi‑
cians, when employed in meafuring the degrees of the
meridian in Peru, were not fupplied with better barome‑
ters, and that they made not obfervations at correfpond‑
ing times; fince the fcene of their operations was un‑
doubtedly preferable to any other on the furface of the
globe, for determining many curious points with refpect
to the modifications of the atmofphere in the torrid
zone: neverthelefs, by attending diligently to what Mr.
BOUGUER [n] hath told us, of the fteadinefs of the baro‑

meter

(n) He fays, that at the South Sea, REAUMUR's thermometer, in the morn‑
ing before Sun-rifing, ftood at 19°, 20°, or 21°; and in the afternoon at 26°,
27°, or 28°. The refpective means correfpond to 76°¼ and 92°⅘ of FAHREN‑
HEIT, and make the mean heat of the day 84°¼. At Quito the temperature
continued at 14° or 15°, anfwering to 65°⅘ of FAHRENHEIT. At the fummits
of Coraçon and Pichincha, the thermometer ftood in the morning feveral
degrees below freezing, and varied 17° in the heat of the afternoon; whence
the mean temperature at thefe higheft ftations, would probably be about 43°½ of
FAHRENHEIT. He further fays, that in the torrid zone, whatever may be the

mean

meter throughout the year; the uniformity of the mean temperature in every affigned ftation; and his mode of computing, by means of the tables of common logarithms, the altitudes of the Cordillero mountains above the valley that extends itfelf between them; it will be no difficult matter to difcover, nearly at leaft, what fort of equation became neceffary; and what were the relative weights of quickfilver and air of the mean temperature, not only in that high region of the atmofphere, but alfo at the level of the fea.

Thus, by infpecting the table of computations, it will appear, that columns of air, whofe bafes were removed fix or eight thoufand feet from the level of the fea, and whofe heights equalled that diftance, when the temperature was $55°$ of FAHRENHEIT, as determined from the mean between the coldeft of the morning and hotteft of the afternoon, the mean logarithmic defect was only $\frac{36.3}{1000}$: whereas, in meafuring heights near the level of the fea, in middle latitudes, the common equation for that temperature is about $\frac{57}{1000}$. The mean ratio of the weight of quickfilver to air, on thefe long columns comprehended refpectively between Carabourou and Quito, and the

mean heat in any affigned ftation, it continues uniformly the fame throughout the year. In this rough eftimation of the temperature in Peru, it feemed unneceffary to examine, whether the true thermometer of REAUMUR was ufed or not; as it could produce no material difference, except at the very hotteft ftations.

fummits

fummits of Pichincha and Coraçon, is that of 16793
to 1. On the altitude of 1534 feet, intercepted between
Carabourou and Quito, which fhort fection of the column is about half-way between the level of the fea, and
the fummits of the Cordilleros, the mean temperature
being 66°, the ratio is that of 15089 to 1 : hence it
feems probable, that quickfilver would have to the different fections of the general column of air, comprehended between the fea and the top of Coraçon, nearly
the following ratios :

	Temp.	
At the level of the South Sea,	$84\frac{1}{2}$	13100 to 1
Half-way from thence to Carabourou,	$75\frac{1}{2}$	14100
At Carabourou, — —	$66\frac{1}{2}$	15100
Half-way from thence to Coraçon,	55	16100
At the fummit of Coraçon, —	$43\frac{1}{2}$	17100
Whereof the mean is, —	65	15100

Mr. BOUGUER tells us, that the barometer in the torrid zone varies not at the fea-fhore above two and a half,
or at moft three lines throughout the whole year. At
Popayan, its variation is only a line and a half; and at
Quito a fingle line. Now let us fuppofe, that an altitude
had been meafured with the barometer at the level of
the

the South-fea, where the defcent of quickfilver at the
upper ftation was exactly an inch in the mean heat of
the day, anfwering to $84°\frac{1}{2}$. On the former fuppofition
of the weight of.quickfilver to air, the height would be
13100 inches or 1091.7 feet.

$$\text{Hence} \begin{Bmatrix} 29.930 & 84\frac{1}{2} - 169 = 29.761 \\ 28.930 & 84\frac{1}{2} - 169 = 28.761 \end{Bmatrix} = 890.6 \text{ feet;}$$

the logarithmic refult, which is defective 201.1, or
nearly $\frac{226}{1000}$ parts. Now this equation being divided by
2.45 the mean expanfion of air, we have nearly 92°
for the difference between 84°$\frac{1}{2}$, the temperature of the
obfervation, and the zero of the.fcale, which reduces it
to $-7°\frac{1}{2}$ of FAHRENHEIT. If it fhould be thought that I
have fuppofed the air to be too light at the level of the
fea under the equator, let it be taken to quickfilver only,
at a mean between 13100 and 12672, which feems to
have been the ratio of their weights at Genoa, when Mr.
DE LUC's temperature was 79°, and we fhall have
12881 inches, or 1073.4 feet of air, for the counter-
poife to the inch of quickfilver in the barometer: hence
$1073.4 - 890.6 = \frac{181.8}{890.6} = \frac{200}{2.45} = 83°.7$, will denote the number
of degrees that the zero of the fcale would, in that cafe, be
below the temperature of the air, which brings it to within
lefs than a degree of the cypher of FAHRENHEIT. But in
middle latitudes the zero of the fcale is at 32°, and the

4   equation

equation, applicable at the level of the fea for the heat of $84°\frac{1}{7}$, is at moft only $\frac{132}{1000}$ inftead of $\frac{205}{1000}$.

Mr. BOUGUER found, that the rule which his expe-rience had furnifhed him with, for computing heights with the barometer between the ranges of the Cordil-leros, namely, that of deducting $\frac{1}{30}$th part from the number of toifes expreffed by the logarithmic dif-ferences, which agrees nearly with the equation $\frac{363}{1000}$ which I have made ufe of in the table of computations, would not anfwer when he came to apply it at the level of the fea. He tells us, indeed, that the elafticities of the air, above and below, are there, as well as in Europe, exactly pro-portionable to its condenfations; and even, that the in-tenfity of the elaftic force, or fpring of the air, is every where equal in all places of the torrid zone that are con-fiderably elevated. The real condenfations in each place are proportionable to the weights of the fuperior columns caufing the compreffion; thefe condenfations being in geometrical, the heights are in arithmetical progreffion. But below the fame law doth not take place; becaufe the intenfity of the elaftic force is really confiderably lefs at the level of the fea, than it is at one or two hundred toifes above it, notwithftanding the effect of the heat, which fhould render it greater. It is to be obferved, that Mr. BOUGUER hath not given us the obfervations whereon

he

he founded this laft deduction; and his note on the text, which I apprehend, neverthelefs, conveys his true meaning, is contradictory to it: for there he fays, that the dilatation occafioned by the heat throughout the day, changed the diftribution of the weight with regard to all the places fituated within the Cordilleros, as well as on other mountains, and made the lower fections of the columns contain lefs and the upper fections more air, than they fhould have done, had it been a perfectly elaftic fluid.

Having now mentioned all the barometrical obfervations that have come to my knowledge, tending any way to throw light on this very intricate fubject, it remains to fum up, from the whole, the general principles whereon I have proceeded in conftructing the table of equation for the heat of the air.

It will be remembered, that I have more than once remarked, that in the Britifh obfervations, when the tem—perature was $52°$, the defect was $\frac{49 \text{ or } 50}{1000}$, the lowermoft barometer ftanding at or near the level of the fea; but in the obfervations on Tinto, a confiderable hill appertaining to the third clafs, whofe bafe is elevated 700 feet above the level of the Clyde at Glafgow, when the temperature was $52°$, I found the equation to be little more than $\frac{41}{1000}$. Again, thefe two facts being compared with

2   the

the aggregate refult of Mr. DE LUC's obfervations, where the lowermoft barometer ftood about 1300 feet above the fea, the equation for the fame temperature feemed not to exceed $\frac{41}{1000}$. Laftly, thefe circumftances being confronted with the refults of Mr. BOUGUER's obfervations, where the lowermoft barometer ftood from 6000 to 8000 feet above the fea, the mean equation for 55° was only $\frac{16.1}{1000}$, which gives $\frac{14}{1000}$ for the heat of 52°. Now thefe Peruvian obfervations, which I believe to be exceedingly good from the fteadinefs of the barometer in that part of the world, being fubftituted in lieu of thofe not yet obtained in our own quarter of the globe, there feemed to me to be a neceffity for concluding, that the equation for middle latitudes, with any affigned temperature above or below the zero of the fcale, diminifhed as the height of the place above the fea increafed; which confequently implied, that the magnitude of the logarithmic terms increafed fafter than the dilatations of the air. But when the comparifon was carried yet farther, and the obfervations in Peru and at Spitzbergen were fairly brought into one view, there appeared to be fufficient grounds for fufpecting, if not abfolutely for concluding, that there could be no fixed zero for the fcale depending on the temperature of the air; but that it would change with the denfity of the atmofphere appertaining to the

latitudes,

latitudes, climates, or zones of the earth, where the obfervations were made. On this fuppofition it was natural
for the mind to form to itfelf fome general hypothefis,
which might ferve to account for the appearances; and
the firft that prefented itfelf was the following: that the
atmofphere furrounding our globe might poffibly be
compofed of particles, whofe fpecific gravities were
really different; that the lighteft were placed at the
equator; and that the denfity of the others gradually increafed from thence towards the poles, where the hea-
vieft of all had their pofition [o]

It is a well known and eftablifhed fact, that in the
middle latitudes, a North or North-eaft wind conftantly
raifes the barometer, and generally higher as its continuance is longer. The contrary happens when a South
or South-weft wind blows; for I believe it is commonly
loweft when the duration and ftrength of the wind from

[o]. It was fuggefted by Dr. GEORGE FORDYCE, that equatorial and Green-
land air might be brought bottled up, and weighed in this country in air of the
refpective temperatures, by means of a curious balance whereof he is poffeffed,
in order to fee whether any difference could be difcovered in their fpecific
gravities. A thought of the fame kind, but more eafily put to experiment,
occurred to Lieutenant GLENIE, of the Royal Artillery, namely, that of
weighing equatorial and polar fea-water. To this gentleman I am obliged for
his affiftance in part of the manometrical experiments, as well as in feveral of
the computations.

that quarter have been the greateſt. Thus the North-eaſt wind, by blowing for any length of time, brings into the middle latitudes a maſs of air heavier than that which naturally appertains to the region, and raiſes the barometer above its mean height. The continuance of a South-weſter carries off the heavy air, depoſits a much lighter body in its ſtead, and never fails to ſink the barometer below its mean height: hence, in the middle parts of Europe, there is a difference of about two inches and a quarter between the higheſt and loweſt ſtates of the barometer. But ſuppoſing it to be only two inches, the difference of preſſure ſtill amounts to $\frac{1}{15}$th part of the whole weight of the atmoſphere. Now it is evident from the Peruvian obſervations, that the greateſt fluctuation of the barometer, which is at the level of the ſea, doth not exceed 0.226 of an inch, or $\frac{1}{133}$d part of the whole preſſure; and if the change ſhould be no greater at the poles, which I think not improbable, it follows, that the meaſurement of heights by means of the barometer, in middle latitudes, will be more precarious and uncertain than in the torrid and frigid zones.

Such in general were the firſt ideas which the compariſon of the operations of the barometer with the effects

of

of the North-eaft and South-weft wind [p] on that inftrument, fuggefted with regard to the different denfities of the atmofphere in the different zones of the earth.

But fince the experiments on the expanfion of moift air have fhewn its elafticity to be fo much greater than that which is dry, I apprehend, that the fimple principle of heat and moifture may fuffice to account for all the phenomena. Thus it is univerfally admitted, that there is a greater degree of humidity and heat in the air, near the earth's furface, than there is in the higher regions of the atmofphere. The elafticity or expanfion of the lowermoft fection [q] of every column of air, whether long

(p) I have been well informed, that in China the North-weft wind raifes the barometer moft, and is highly electrical; it is at the fame time the drieft and the coldeft; and at Canton, under the Northern tropic, there is frequently ice. On the Eaft-coafts of North America the feverity of the North-weft wind is univerfally remarked; and there can fcarcely be a doubt, that the inhabitants of California, and other parts on the Weft-fide of that great Continent, will, like thofe on the Weft of Europe, feel the ftrong effects of a North-eaft wind. The extraordinary drynefs and denfity of the wind from the North pole, feems therefore to be occafioned by its paffing over the Continent of Europe and Afia on one fide, and that of North America on the other. Thofe who live on the Eaft and Weft-coafts of South America, will find the drieft and coldeft winds come to them refpectively from the South-weft and South-eaft. As the winds feem to be colder, drier, and denfer, in proportion to the extent of land they pafs over from the poles towards the equator, fo they appear to be more moift, warm, and light, in proportion to the extent of Ocean they pafs over from the equator towards the poles. Hence the humidity, warmth, and lightnefs, of the Atlantic wind to the inhabitants of Europe.

(q) Mr. DE LUC feems to have fufpected fomething of this kind towards the

5 D 2   end

long or fhort, will confequently be greater than the uppermoft fection of it; for the heat, by diffolving the
moifture, produces a vapour lighter than air, which mixing with its particles, removes them farther from each
other, increafes the elafticity of the general mafs, and
diminifhes its fpecific gravity comparatively more than
it doth that of the fection immediately above it, where
there is lefs heat and lefs moifture. Hence I infer, that
the equation for the air, in any affigned vertical, will
gradually diminifh as the elevation of the place above
the fea increafes, and that it will vanifh at the top of the
atmofphere. This is in fome refpect confirmed by the
experiments on the expanfion of rare air; for from them
it appears, when the particles are very far removed
from each other, by a great diminution of preffure, as is
undoubtedly the cafe in the higher regions of the atmofphere, they lofe a great part of their elaftic force. Thus
the equation, anfwering to any particular temperature,
above or below the zero of the fcale, at different heights
above the furface, will, I apprehend, be expreffed by the
ordinates to a curve of the hyperbolic order, whofe cur-

end of his 8th chap. *fur les difficultés à vaincre:* and in that which follows, he
gives proofs of the lightnefs of vapours with regard to air, faying, that they
point out fire to be their vehicle. He afterwards quotes NEWTON with refpect
to the lightnefs of a humid atmofphere compared with one that is dry.

5   vature

vature may be fuppofed to change faft near the furface of the earth, and differ infenfibly from a ftraight line at great heights above it.

With regard to the latitudinal equation, the fame principle of heat and moifture feems to make it probable, that fuch will become neceffary in operating with the barometer; for it is well known, that there is a great degree of humidity in the air between the tropics; and, on the contrary, that the polar atmofpheres are very dry. The heat and moifture being greateft at the equator, there the elafticity or equation will likewife be the greateft at the level of the fea; and the zero of the fcale will neceffarily defcend to a lower point of the thermometer, than that to which it correfponds in middle latitudes. As the elafticity of the air at the level of the fea, or equal heights above it, with the fame degree of heat, will always be proportionable to the quantity of moifture diffolved in it, therefore it will gradually diminifh from the equator towards the poles; that is to fay, the zero of the fcale will afcend in the thermometer, coincide with the 32d degree in the middle latitudes, and, in its motion upwards, will give the equation to be applied with the contrary fign in high latitudes. Hence I infer, that every latitude, climate, or zone, will not only have its particular zero, but alfo its particular curve, whofe ordinates

will

will always meafure the equations applicable in the refpective fituations. The equatorial curve will probably
change the fafteft, and the others become gradually flat-
ter, as they approach towards the poles, where the
greater, but more uniform, denfity of the atmofphere
may occafion it to differ little from a ftraight line. I apprehend, however, that even at the pole fome fmall diminution might be found to take place in the equation,
was it poffible, in that region, to prove it by experiments
at a fufficient height above the level of the fea.

The table of the equation, depending on the heat of
the air, annexed to this paper, is conftructed for middle
latitudes. It extends to temperatures from 12° to 92° of
FAHRENHEIT; and for fituations fo greatly elevated
above the fea, as to make the mean barometer between
the two ftations ftand no higher than 19 inches. As the
equation correfponding to the lower parts of the atmofphere, contained in the right-hand columns, will come
more frequently into ufe than that appertaining to the
higher regions, comprehended in thofe on the left;
therefore, in the firft, it is given for every half; and in
the laft only, for every whole inch of defcent of quickfilver in the tube.

The equation found in the column of 29 inches, correfponds exactly with the expanfion of air refulting from

I   the

the manometrical experiments; and the ratio of diminution, in the temperature of 52°, hath been taken from
the Peruvian obfervations, fuppofing it to decreafe uni-
formly $\frac{1}{1000}$ on each inch, or $\frac{16}{1000}$ on the eight inches between 29 and 21. For the fake of fimplicity, as well as
from the want of fufficient data for afcertaining the
lengths of the ordinates of the curve, the arithmetical
hath been preferred to any progreffive diminution that
might have been adopted, though by this mode the refults would have agreed better with fome of my own, as
well as Mr. DE LUC's obfervations. In each of the columns the equations for particular temperatures, compared with that for 12° or 92°, are reciprocally proportionable, fo that the maximum of the rate always correfponds to the fpace between 52° and 72°, as indicated
by the manometer. It will be obferved, that though the
equation in the table is only given for every 10° of difference of temperature, yet, by the intermediate rates for
fingle degrees in the columns refpectively, and the ratio
of diminution for the height of the mean barometer
above the fea, expreffed in that towards the right-hand,
the equation for any particular temperature may be readily obtained. The application of this table makes the
third part of the rule, for meafuring heights with the
barometer. When the mean temperature of the column

of

of air is above 32° of FAHRENHEIT's thermometer in the
fhade, add the equation correfponding to the temperature and height of quickfilver in the mean barometer to
the logarithmic altitude; when below 32°, fubtract the
equation from the logarithmic altitude; the fum in the
firft cafe, and difference in the laft, gives the real height.

Befides the table of equation for the air, adapted to the
meafurement of the greateft acceffible heights the barometer can poffibly be applied to in middle latitudes, I
have annexed, for the ufe of thofe who may prefer fimplicity, and ftill doubt of the vertical diminution, a thermometrical fcale of the equation, fuited to Englifh and
French meafures, with their refpective thermometers. It
will readily be conceived, that the divifions, expreffing
the 1000th parts in this fcale (r), are unequal, fince they
follow the inverfe ratio of the thermometrical compared
with the manometrical degrees. Where thefe laft are
the greateft, as between 52° and 72°, the divifions expreffing the equation are the fmalleft, becaufe a greater

(r) Any fcale of this kind, unlefs it had been mechanically divided by a
mathematical inftrument-maker, could not be rendered very exact; and it may
be expected, that the imperfections in the original will be augmented in copy-
ing by the engraver, notwithftanding the utmoft care on his part : wherefore,
on the left-hand fide of the plate, I have annexed the number of degrees and
decimal parts of FAHRENHEIT, below the temperature of 91°.88, correfponding
to every $\frac{1}{1000}$th parts of the equation, by which means the unequal fcale may,
at any time, be divided with fufficient accuracy.

number

number of them correſpond to the ſame thermometrical
ſpace. When the height is required in fathoms, the
zero of FAHRENHEIT correſponds to −71.72, and the
boiling point to +412.49: the ſum of the two equations 484.21 is the actual expanſion of common air from
the heat of 212°. When the French toiſe is made uſe
of as the meaſure, the zero of the ſcale hath been ſhewn
to coincide with 57°.18 of FAHRENHEIT, or +11°¼ of
REAUMUR. The negative equation 134°.72 anſwering
to −14°¼ of REAUMUR, and the poſitive 349°.49 correſponding to +80°, or the boiling point, being added
together, make again 484.21.

In order to convey a more diſtinct idea of the effect
which heat produces in the dilatation of different kinds
of air, compared with quickſilver, along with the ſcale
for the equation I have placed another, expreſſing the
actual and relative expanſions, reſulting from the mean
of the experiments, for every 20° of difference of temperature. This ſcale is intended to give a comparative
view of the manometrical with the thermometrical
ſpaces, mentioned in the ſecond ſection.

I ſhall now cloſe this paper, which hath already
greatly exceeded the limits I wiſhed to have been able to
preſcribe to it, with a few remarks on the error of the rule,
perceivable in the tables of computation, and the mea-

fures that fhould, in my opinion, be taken to bring the
theory of the barometer to a ftill greater degree of perfection, fuch as I believe it to be really capable of.

By infpection of the tables containing the computations of the Britifh obfervations, it will be feen, that the
error of the rule is in general very fmall. In the London clafs it is greateft on Shooter's-hill, making the
height five feet too little. In thofe at Taybridge, one of
the obfervations on Schihallien gives a defective refult of
$29\frac{1}{2}$ feet; but this is eafily accounted for, as it certainly
arofe from the fhort time given to the barometer to lofe
the heat it had acquired in carrying upwards, thofe
deftined to obferve at the fummit arriving there too late,
that is to fay, towards the expiration of the whole hour
which the inferior barometer had been obferved in. One
of the obfervations on Carmichael-hill, though a fmall
height, is defective eight feet, which I afcribe to the
South-weft wind and humidity of the air. From the
fame caufe I would account for the uniform defect in
the firft part of the Edinburgh obfervations: in the laft
part of thefe, the circumftances having changed, the error hath the contrary fign. In the Linhoufe clafs of obfervations, the fame caufe of defect appears on the 1ft
of December, 1775, and on the 20th of November,
and 17th of December, 1776. The only refult which I
                                        confider

confider as very irregular, and do not pretend to account for, is that for the height of Moel Eilio, a hill fituated between Carnarvon and Snowdon: the real altitude 2371 feet, is exceeded by the barometrical refult 21 feet, though the circumftances were fuch as, in other cafes, generally make it fall fhort.

At the bottom of the tables of computations I have occafionally fubftituted Mr. DE LUC's equation for the air, in calculating one or more of the greateft heights, that the difference between the two methods might become more obvious. Thus the firft obfervation on Schihallien is defective $67\frac{1}{2}$ feet; the mean of thofe on Tinto 29 feet; Moel Eilio 41 feet; and Snowdon 81 feet.

With refpect to the refults that the rule produces on Mr. DE LUC's heights it will be obferved, that it anfwers very well in the cold obfervations, which, with his rule, were often defective 60 or 70 feet; but gives too much in thofe that are hot. If, however, the whole of thefe hot obfervations had been included, the apparent error would have been lefs; for the mean defect was taken at $\frac{96}{1000}$ for the hotteft temperature, whereas it fometimes amounts to $\frac{110}{1000}$. On the height of the Dole the rule errs in defect; and on the mean of Mr. DE LA CAILLE's obfervations, at the Cape of Good Hope, it exceeds the truth. By fubftituting Mr. DE LUC's equation for the air,

5 E 2

in

in the computations of the Dole and Table-hill, the refpective refults are defective 96 and 62.6 feet.

To the Britifh obfervations a table is annexed, containing the barometrical computations of altitudes not yet
determined geometrically. In the chief part of thefe the
inferior barometer ftood at Belmont-caftle, the feat of the
lord privy-feal for Scotland, by whofe directions the correfponding obfervations were made. This table likewife
comprehends Mr. BANKS's obfervations in 1772, for the
height of the South-pap of Jura, above Freeport in the
ifland of Ifla, and thofe he made the fame year, to obtain the height of Mount Hecla, above Hafniford in
Iceland.

Laftly, it is to be obferved, that in the application of
the table, the equation found in the columns $29\frac{1}{2}$, 30,
and $30\frac{1}{2}$, will never come into ufe, except in the meafurement of fhort columns of air, whofe bafes ftand at,
or not much above, the level of the fea[1]. In an ifland,

whofe

(1) Having been accuftomed, from the beginning, to call the ftation of the
inferior barometer the place of obfervation, and to fuppofe the mean height of
its quickfilver to denote the elevation of the place above the fea, for the fake
of fimplicity I adapted the formula to the height of quickfilver in that barometer, and made all the computations in the tables accordingly. But it having
been fuggefted to me, firft by Sir GEORGE SHUCKBURGH, and afterwards by
Mr. DE LUC, that this mode, though the eafieft, was not ftrictly accurate, nor
confiftent with the principles whereon a vertical diminution of the equation for

the

whofe climate is fo very variable as that of Britain, fet-
tled weather fhould be chofen as the beft time for obfervations. With any fudden fall of the barometer, in any
affigned ftation below its mean height, it is apprehended
that the rule will have a tendency to give defective refults; and the contrary fhould happen when, from the
increafed weight of the atmofphere, it rifes much above
the mean height.

From what hath been faid in the courfe of this paper,
it will be perceived, that though the error of the rule is
in general very fmall, yet now and then fuch irregularities do occur as plainly fhew, that fomething ftill remains
to be done, in order to perfect the theory of the barometer.

The exiftence, or otherwife, of a latitudinal equation
being a point of the greateft confequence, fhould be determined with fo much care as to leave no doubt remaining on that head. And as this can only be effected by
differences that are extremely obvious, the obfervations
for that purpofe fhould be made at the equator, and as
near as poffible to the poles[t]. Peru is no doubt the beft
fituation

the heat of the air was founded, I have fince changed it to the mean barometer,
or middle of the column of air intercepted between the two ftations. In this
way all the great heights have been re-computed: the fmaller altitudes, not
being fenfibly affected by the alteration, continue as at firft.

(t) Some idea may be formed what altitudes on the furface of the globe are
acceffible

fituation on the globe for conclufive equatorial obfervations; but as it would be found very difficult to carry any
fcheme of that kind into execution, fuch as may be more
eafily obtained in our Weft India iflands, which have the
higheft mountains, would be very fatisfactory with refpect to the expanfion and weight of moift air, at different heights above the furface. At the tops of the
mountains in the torrid zone, the obfervations would
always be fufficiently cold; but it would be of ufe likewife, to have the coldeft poffible at the level of the fea,
under or near one of the tropics, when the Sun was in
the other.

With regard to obfervations in the frigid zones, Spitzbergen feems to be as proper a fituation as any; though
others may no doubt be found in the Northern parts of
the Ruffian empire: and it is prefumed, that the Pe-
terfburg academy would direct the experiments to be
made.

accefiible to man, by confidering the height above the fea of the inferior line
of perpetual fnow. In the middle of the torrid zone it appears, from Mr.
BOUGUER's obfervations, to be elevated 5201 yards, and 4476 about the tropics. In middle latitudes there is everlafting fnow on the mountains at the
height of 3300 yards. In the latitude of 80° North, Lord MULGRAVE found
the inferior line of fnow to be only about 400 yards above the fea: whence we
may conclude, that the furface of the earth, at the pole itfelf, is for ever
covered with fnow.

The Peak of Teneriffe, Ætna, the mountains of Auvergne and Rouffillon, as well as Hecla in Iceland, are all very proper for obfervations in intermediate latitudes.

Within the ifland of Great Britain, Ben Nevis feems to be the beft mountain for barometrical obfervations, becaufe of its great height, its vicinity to the fea, and that there is very good ground clofe to its foot (which is rarely the cafe in the Highlands) for the meafurement of the bafe that would be made ufe of in the geometrical operations.

One of the chief caufes of error in barometrical computations, I apprehend, proceeds from the mode (though fimplicity is in its favour) of eftimating the temperature of the column of air from that of its extremities, which muft be faulty [u] in proportion as the height and difference of temperature are great. Where very accurate conclufions are expected, fimultaneous obfervations, at different times of the day, and different feafons of the year, fhould be made with feveral barometers, placed at different heights, each accompanied with a thermometer and manometer. By this method, the progreffion of temperature, as well as the law of diminution of the equation, from the pofition of the inferior barometer above the fea (if fuch diminution doth really take place)

(u) This is taken notice of by Mr. DE LUC.

would

would þe obtained with certainty. Suppofing, for inftance, Ben Nevis was divided into four fections, five barometers, with as many obfervers, would be neceffary. This number may feem great, but the expence of people employed in that way would be inconfiderable. And if it fhould be judged proper, there could not furely be any great difficulty in providing reafonable accommodation for an obferver, who fhould live a whole year at the top of the mountain, while another made correfponding obfervations below.

But the perfecting of the theory of the barometer is not the only advantage that would accrue from a combination of thefe obfervations; for, while they were carrying on in different climates, or zones of the earth, good opportunities would offer of determining the refractions, as well as the force of gravity and figure of the globe, from the vibrations of the pendulum.

The mean expanfion of common air is already found to be greater than what was formerly fuppofed; wherefore the mean refraction will be altered proportionably. And fince the expanfion of moift air is found to be fo much greater than that of common air, a larger field for inquiry and inveftigation is now laid open.

With refpect to the experiments with the pendulum, Mr. BOUGUER feems to have been the only perfon, fo far

as

as I know, who hath taken the denfity of the medium
in which it performed its vibrations into the account, and
given us its length at the equator in vacuo. But if we
are to judge of the denfity of the air in the frigid zone
from the barometrical obfervations at Spitzbergen, the
pendulum there muft have loft fo much of its weight, as
to have leffened confiderably the number of vibrations
below what they would have been in vacuo, in the fame
temperature. Having confidered the effect that this
would produce, I collected the beft experiments that
have hitherto been made with the pendulum into one
view, and having applied the equation that the denfity
of the air, in which they feverally vibrated, feemed to
require; I found from computation, that the ratio of the
diameters of the earth is (as Mr. BOUGUER fuppofed it)
nearly that of 178 to 179, inftead of 229 to 230, as
eftimated by Sir ISAAC NEWTON, and which agrees very
nearly with the mean refult from the meafurement of
the degrees of the meridian. The experiments with the
pendulum are fo fimple and eafy, may be repeated fo
often in all fituations, and are fo much more confiftent
with each other, than the meafured lengths of degrees
of latitude, that it appears to be incomparably the beft
method for determining the figure of the earth. And if
it fhould really be found fo flat a fpheroid as the pendu-

lum

lum feems to make it, both parallaxes and refractions, will require correction.

Upon the whole, though I wifhed to be concife in the recital of the experiments and obfervations contained in this paper, yet I found it neceffary at the fame time to be explicit. Some of them were either entirely new, or managed in a different manner from what they had formerly been. This forced me into a comparifon of many minute circumftances attending the operations, and to a tedious, though neceffary, combination of the various refults. Without taking a comprehenfive view of the whole matter, and ftating every thing with fairnefs and candour, I could not convey to others the ideas I entertained of it myfelf; nor enable them to judge, how far I had been juft in the conclufions already drawn, or confiftent in my fuppofitions concerning fuch points as are yet left doubtful. If I have been obliged to differ from Mr. de luc, it is becaufe the Britifh obfervations, as well as his own (confidered by their extremes) feem to authorize it: he is himfelf too candid to fuppofe, that I have had criticifm in view, or indeed any other object, than that of contributing my mite towards the difcovery of the truth, from the very good foundation which he hath already laid for it. I am aware it may be alledged, that I have rendered the theory of meafuring heights by

5   the

the barometer fo much more complicate and difficult, as perhaps to deter others from applying it to ufeful purpofes. To this I anfwer, that though it feem utterly impoffible to render what is really intricate in its nature, extremely fimple; yet that the beft and fureft method of arriving at fimplicity at laft will, in the firft place, be to afcertain the limits of deviation of the rule, by a proper number of good obfervations, made in circumftances and fituations as different as poffible from each other. In the prefent ftate of the matter, I doubt not but the barometer will be found to give refults fufficiently near the truth for all ordinary purpofes, the nicer bufinefs of levelling alone excepted. It is the only inftrument by which the relative heights of places, in very great and diftant tracts of country, can eafily and fpeedily be obtained, by the preffure of the atmofphere alone. The method of ufing it is attainable by all, requiring only a little habit, and fome degree of attention to prevent the admiffion of air into the tube. Few people are qualified for the tedious and very laborious operations of accurate geometrical meafurements. Mountainous countries rarely afford bafes of fufficient length, which, to avoid error, muft be meafured again and again with the utmoft care. Inftruments of the moft expenfive kinds muft be employed to take the angles; at the fame time that a thorough knowledge of their ufe,

and

and a fcrupulous attention to their various adjuftments, become indifpenfably neceffary. In fhort, the facility of one method, compared with the other, is fo exceedingly obvious as to need nothing elfe to recommend it as a fubject very curious and ufeful, and therefore well worthy of the refearches of philofophers, till, by their united labours, it hath been brought to perfection.

N° I.

Table fhewing the equation depending on the temperature of the column o

inferior

Mean temperature of the column of air.		Inches 19		20	21	22		23	24	25		26		26¼	
						Mean equated height of quickfilver in the inferior and fuperior ba									
92°	Add to the logarithmic altitude.	89.364	1.44	95.456	101.548	107.640	1.73	113.733	119.825	125.918	2.03	132.010	2.13	135.057	2.
82		74.967	1.49	80.078	85.189	90.300	1.80	95.411	100.522	105.633	2.10	110.744	2.21	113.299	2.
72		60.028	1.52	64.120	68.213	72.305	1.83	76.398	80.491	84.583	2.14	88.675	2.25	90.722	
62		44.818	1.55	47.873	50.928	53.983	1.87	57.039	60.094	63.150	2.18	66.205	2.29	67.733	
52		29.335	1.49	31.335	33.335	35.355	1.80	37.335	39.335	41.335	2.10	43.335	2.21	44.335	2
42		14.394	1.44	15.376	16.358	17.340	1.73	18.321	19.303	20.284	2.03	21.266	2.13	21.757	2
32							When the mean temperature of the column of air to be mea								
22	Subtract from the log. altitude.	13.852	1.39	14.796	15.741	16.685	1.67	17.630	18.575	19.519	1.95	20.463	2.05	20.936	2.
12		27.162	1.33	29.014	30.866	32.718	1.60	34.569	36.421	38.273	1.88	40.125	1.97	41.051	

ımn of air, and its elevation above the fea, as denoted by the mean height of ferior and fuperior barometers.

perior barometers. Equation in thoufandth parts of the logarithmic altitude.

26½		27		27½		28		28½		29		29½		30		3
											2.38					
5.057	—	138.103	—	141.149	—	144.195	—	147.242	—	150.288	—	153.334	—	156.381	—	159.
	2.18		2.22		2.27		2.32		2.37		2.42		2.47		2.52	
3.299	—	115.855	—	118.411	—	120.966	—	123.522	—	126.077	—	128.632	—	131.188	—	133.
	2.26		2.31		2.36		2.41		2.46		2.51		2.56		2.61	
0.722	—	92.768	—	94.814	—	96.860	—	98.907	—	100.953	—	102.999	—	105.047	—	107
	2.30		2.35		2.40		2.45		2.51		2.56		2.61		2.66	
7.733	—	69.261	—	70.789	—	72.316	—	73.844	—	75.372	—	76.900	—	78.427	—	79.
	2.34		2.39		2.45		2.50		2.55		2.60		2.66		2.71	
4.335	—	45.335	—	46.335	—	47.335	—	48.335	—	49.335	—	50.335	—	51.335	—	52.
	2.26		2.31		2.36		2.41		2.46		2.51		2.56		2.61	
1.757	—	22.248	—	22.739	—	23.229	—	23.720	—	24.211	—	24.702	—	25.193	—	25.
	2.18		2.22		2.27		2.32		2.37		2.42		2.47		2.52	

o be meafured is at 32°, the differences of the logarithms give the real height in fathoms and 1000th parts.

	30½	Rate of diminution for whole and half inches.	
—		—	⎧ 6.0925
ı.52	159.427	2.57	⎩ 3.04625
—		—	⎧ 5.111
..61	ı33.743	2.66	⎩ 2.5555
—		—	⎧ 4.09250
..66	107 093	2 71	⎩ 2 04625
—		—	⎧ 3.0555
.71	79.954	2.76	⎩ 1.52775
—		—	⎧ 2.000000
.61	52.335	2.66	⎩ 1.000000
—		—	⎧ 0.981625
.52	25.684	2.57	⎩ 0.490812

Add to the logarithmic altitude.

.42		2.47	⎧ 0.47225
—	24.714		⎩ 0.94450
.33		2.37	⎧ 0.925875
—	48.458		⎩ 1.85175

Subtract from the log. altitude.

I. Compu-

SCALE for the Equation of the Air.

English Measure   French Measure

Equation in some part of the Logar: Altitude.	Fahrenheits Thermom:	Equation in some parts of the Log: the Altitude.	Reaumurs Thermom:

$91.88.$

4.15 $87.73.$

4.10 $83.63.$

4.06 $79.57.$

4.02 $75.55.$

$.3.97$ $71.58.$

3.93 $67.65.$

3.87 $63.78.$

3.84 $59.94.$

3.82 $56.12.$

3.88 $52.24.$

3.94 48.30

3.99 $44.31.$

4.05

Add to the Logarithmic Altitude.

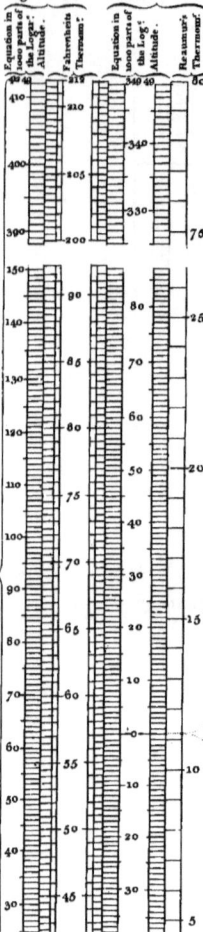

Add to the Logarithmic Altitude.

A SCALE expressing the Expansion of different kinds of Air with regard to Mercury affected by 212° of Fahrenheit.

Actual Expansion of 1000 equal parts	Mercury. 17.06	Common Air. 484.21	Very Rare Air. 166.66	Moift Air. 1038.06

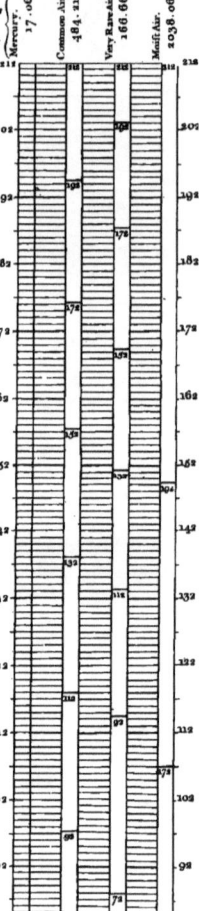

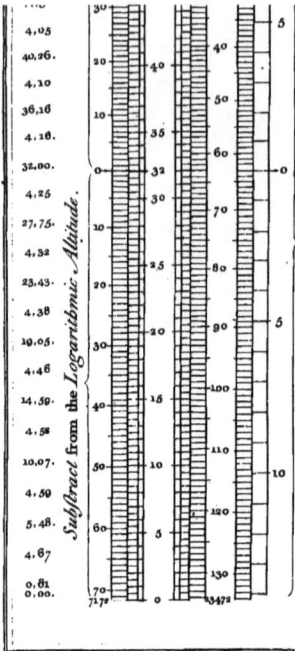
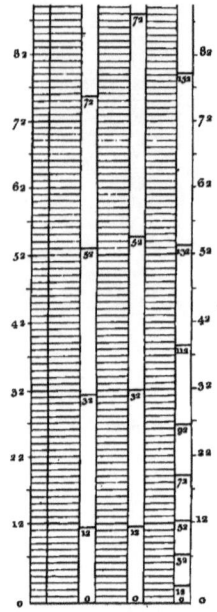

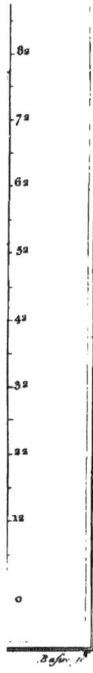

Geometrical heights f the ftations in feet.	Date of the obfervations, winds, &c.	Obferved heights of the inferior and fuperior barometers.	Temperature of the quickfilver.	Equation for the heat of the quickfilver.	Equated heights of the barometers.	Logarithmic refult in feet.	Logar. excefs or defect in ft. and alfo in 1000th parts.	Temperature of the air, Particular.	Temperature of the air, Mean.	Equation by the
t. Paul's Church-yard North-fide, and iron-gallery over the dome, 281 feet.	1774, Dec. 1. 9ʰ 27′ wind N.W.	29.659 29.338	33°¾ 34	—006 —006	29.653 29.332	283.6	+2.6ft. = 9/1000	32°¾ 33¾	33°¼	+ =
	Dec. 31. 11ʰ 52′ A.M. wind N.E.	30.187 29.864	35½ 34¾	—011 —009	30.176 29.855	279.5	— 1.5 = 5.5	33 34	33½	+ =
	April 22. 11ʰ 55′ A.M. E. wind.	30.136 29.839	50¼ 53	—060 —069	30.076 29.770	266.5	—14.5 =54.4	49 53	51	+ =
Top of Paul's-ftairs, and the faid gallery, 324 feet.	1774, Ap. 22. 0ʰ 50′ P.M.	30.206 29.842	55¼ 53¾	—080 —071	30.126 29.771	308.9	—15.1 =49.	55¼ 53¾	54½	+ =
	Dec. 1. 10ʰ 2′ A.M.	29.717 29.344	37 35	—016 —009	29.701 29.335	323.1	— 0.9 = 3.	37¼ 34	35⅝	+ =
	Dec. 31. 0ʰ 22′ P.M.	30.230 29.858	35¼ 34¾	—011 —009	30.219 29.849	321.	— 3. = 9.3	34¾ 34¼	34½	+ =
Scotland-yard wharf, and Old Spaniard dining-room, 422 feet.	1774, Dec. 24. 10ʰ 7′ A.M. N.E. wind.	30.844 30.349	33½ 33¼	—005 —004	30.839 30.345	420.8	— 1.2 = 3.5	34¼ 33¼	33¾	+ =
Great Pulteney-ftreet, and the faid dining-room, 352 feet.	1774, Nov. 28. 9ʰ 48′ A.M.	29.684 29.287	35½ 34¾	—011 —009	29.673 29.278	349.2	— 2.8 = 8.5	35¼ 34¾	35	- =
	Dec. 9. 0ʰ 15′ P.M. N.W. wind, fnow.	29.647 29.234	27½ 25¼	+015 +022	29.662 29.256	359.	+ 7. =19.7	27½ 23¼	25½	- :
	Dec. 24. 10ʰ 52′ A.M. N.E. wind.	30.758 30.343	35 33	—010 —003	30.748 30.340	348.1	— 3.9 =11.2	34½ 30¾	32⅝	- :
	1775, June 13. 11ʰ 7′ A.M. S.W. wind.	30.044 29.674	69 69	—121 —117	29.923 29.557	320.7	—31.3 =97.6	67½ 72½	70	· :
	1776, May 10. 10ʰ 30′ A.M.	30.096 29.706	53 51¼	—069 —063	30.027 29.643	335.4	—16.6 =50.	51½ 49¾	50¾	· :
	May 30. 11ʰ 40′ A.M. S.W. wind.	29.900 29.521	66 63	—111 —100	29.789 29.421	323.9	—28.1 =86.7	66 63	64½	· :
	June 20. 0ʰ 15′ P.M.	30.268 29.898	71½ 71½	—129 —127	30.139 29.771	320.1	—31.9 =99.7	71 71½	71¼	· :
	July 16. 0ʰ 15′ P.M.	29.625 29.253	67½ 67½	—113 —112	29.512 29.141	329.6	—22.4 =68.	67¾ 65½	66¾	· :
	Aug. 26. 10ʰ 35′ A.M.	30.132 29.738	59½ 57¾	—092 —082	30.040 29.656	335.2	—16.8 =50.1	59¼ 56½	57¾	· :
	Aug. 27. 11ʰ 45′ A.M.	30.020 29.631	62¼ 60	—099 —091	29.921 29.540	334.	—18. =54.	62 58¼	60	· :

hts in and near London.

ature air,	Equation by the rule in 1000th parts, and also in feet.	Result by the rule,		Error of the rule in feet.	Ratio of the weight of quick-silver to air, air being 1.
Mean.		Particular.	Mean.		
3°¼	$+\frac{3.8}{1000}$ = 1. ft.	284.6	- -	- -	10505
3½	+ 4. = 1.	280.5	281.4	+0.4	
1	+47.5 =12.7	279.2	- -	- -	11056
4½	+55. =17.	325.9	- -	- -	10860
5¾	+ 6.3 = 2.	325.1	324.6	+0.6	
4½	+ 5.8 = 1.8	322.8	- -	- -	10565
3¾	+ 4. = 0.9	421.7	- -	—0.5	10251
5	+ 7. = 2.4	351.6	- -	- -	10694
25½	—14.5 = 5.2	353.8	- -	- -	10404
32⅝	+ 1.2 = 0 4	348.5	- -	- -	10328
70	+96. =30.8	351.5	- -	- -	11416
50¾	+46.9 =15.7	351.1	- -	- -	10971
64½	+81.8 =26.5	350.4	353.5	+1.5	11447
71¼	+101. =32.3	352.4	- -	- -	11416
66¾	+87.4 =28.8	358.4	- -	- -	11355
57¼	+65. =21.8	357.0	- -	- -	10887
60	+70.2 =23.4	357.4	- -	- -	11028

	...ug. 2 ... 45 A.M.	3... 29.631	60	—091	29.540	} 334.	{ =54.	58¼	} 60 {
Pagoda in Kew-gardens, 116.5 feet.	Sept. 2. 10ʰ 15′ AM.	29.294 28.918	60 58¼	—089 —084	29.205 28.834	} 333.1	{ —18.9 =56.8	59¼ 59½	} 59¼ {
	1773, Dec. 20. 1ʰ 22′ P.M. mean of 6 obſervations with 3 barometers.	29.351 29.226	49¼ 49¼	—052 —052	29.299 29.174	} 111.4	{ — 5.1 =45 8	49¼ 49¼	} 49¼ {
Gun-wharf of Woolwich-warren, and upper ſtory of Shooter's-hill inn, 444 feet.	1774. Apr. 27. 4ʰ P.M. mean of 4 obſ.	29.762 29.282	57¼ 56¼	—082 —077	29.680 29.205	} 420.4	{ —23.6 =56.1	52¼ 58¼	} 55¼ {
	Apr. 27. 6ʰ 30′ P.M. mean of 2 obſ.	29.773 29.302	54 55¼	—072 —074	29.701 29.228	} 418.3	{ —25.7 =61.4	49 49½	} 49¼ {
	Apr. 28. 5ʰ A.M. mean of 5 obſ.	29.805 29.336	44¼ 48¼	—041 —053	29 764 29.283	} 424.5	{ —19.5 =46.	43 4 1¼	} 42¼ {

60	{ |=23.4| }	357.4	- -	- -	11028
59½	{ +69. / =23. }	356.1	- -	- -	11355
49¼	{ +43.3 / = 4.8 }	116.2	- -	—0.3	11184
55½	{ +59.2 / =24.9 }	445.3	- -	- -	11170
49¼	{ +43.5 / =18 2 }	436.5	438.9	—5.1	11217
42⅓	{ +24.5 / =16 4 }	434 9	- -	- -	11077

N° II,

Nº II. Computations of barometrical obfervations made on heights ne
and Nº III. of thofe near Lanark.

Nº II. near Taybridge.

Geometrical heights of the ftations in feet.	Date of the obfervations, winds, &c.	Obferved heights of the inferior and fuperior ba- rometers.	Temperature of the quickfilver.	Equation for the heat of the quick- filver.	Equated heights of the barometers.	Logarithmic refult in feet.	Logar. excefs or defect in ft. and alfo in 1000th parts.	Temperature of the air, Parti- cular	Mean.
Station at Weem, and top of Weem- craig, 700¼ feet.	1774, July 16. 11ʰ 30′ A.M. bright Sun-fhine.	29.996 / 29.237	69°¼ / 65¼	—122 / —107	29.874 / 29.130	} 657.2	{ —43.2ft. / = 619/1000	65°¼ / 60 } 62°¼ { :
Ditto ftation, and top of Bolfrack's cairn, 1076¼ feet.	July 16. 6ʰ 30′ P.M. calm and cloudy.	29.933 / 28.788	61¼ / 58¼	—097 / —083	29.836 / 28.705	} 1007.	{ — 69.5 / = 69.	60 / 56¼ } 58¼ { :
Ditto ftation, and top of Dull-craig, 1244¼ feet.	July 18. 7ʰ 20′ A.M.	29.825 / 28.500	58¼ / 55	—086 / —072	29.739 / 28.428	} 1174.	{ — 70.2 / = 60.	57 / 55 } 56 {
Ditto ftation, and top of Knock-farle, 1364¼ feet.	July 18. 5ʰ 4′ A.M.	29.816 / 28.347	55¼ / 51	—077 / —059	29.739 / 28.288	} 1303.5	{ — 61. / = 46.8	54 / 48¼ } 51¼ {
Ditto ftation, and that in Glenmore, 1279¼ feet.	July 12. 7ʰ 30′ P.M.	29.528 / 28.161	58 / 51¼	—084 / —060	29.444 / 28.101	} 1216.5	{ — 62.7 / = 51.6	55 / 51¼ } 53¾ {
Ditto ftation, and South obfervatory on Schihallien, 2098 ft.	July 11. 7ʰ 30′ P.M.	29.643 / 27.432	58¼ / 48	—086 / —048	29.557 / 27.384	} 1989.8	{ —108.2 / = 54.4	58 / 47¼ } 52¼ {
Ditto ftation, and Weft fummit of Schihallien, 3281 feet.	July 11. 7ʰ 30′ A.M.	29.595 / 26.194	59¼ / 46	—089 / —040	29.506 / 26.154	} 3142.3	{ —138.7 / = 44.1	56 / 45 } 50¼ {
Station in Glenmore, and the South obfer- vatory, 818 76.	July 12. 5ʰ A.M.	29.610 / 26.223	50¼ / 44	—062 / —035	29.548 / 26.188	} 3145.5	{ —135.5 / = 43.1	50¼ / 42 } 46¼ {
	July 12. 8ʰ P.M.	28.161 / 27.325	51¼ / 48¼	—060 / —050	28.101 / 27.275	} 777.4	{ — 41.4 / = 53.2	51¼ / 48 } 49¼ {

The obfervation on Schihallien on July 11, by Mr. DE LUC's } 3142.3 50°.5—39°.7=10°.8×2.1 {
equation for the air,

Nº III. near Lanark.

Level of the Clyde at Lanark Bridge, and the ftation at the garden, 362¼ feet.	1774, Aug. 20. 6ʰ 30′ A.M.	29.776 / 29.383	62¼ / 61¼	—099 / —094	29.677 / 29.289	} 342.9	{ —19.6 / =57.1	62 / 62 } 62 { :
	Aug. 23. 3ʰ 8′ P.M.	29.956 / 29.563	64¼ / 65	—107 / —106	29.849 / 29.457	} 344.5	{ —18. / =52.3	63 / 63 } 63 { :
	Sept. 5. 8ʰ A.M.	29.626 / 29.232	52¼ / 50¼	—067 / —060	29.559 / 29.172	} 343.4	{ —19.1 / =55.6	52¼ / 49¼ } 51 { :
	Sept. 7. 7ʰ 47′ A.M.	29.864 / 29.467	50¼ / 51	—061 / —062	29.803 / 29.405	} 350.3	{ —12.2 / =34.8	45 / 44 } 44¼ { :
		29.886 /		—061 /					

rature air, Mean.	Equation by the rule in 1000th parts, and also in feet.	Reſult by the rule, Particular.	Mean.	Error of the rule in feet.	Ratio of the weight of quickſilver to air, air being 1.
62°¾	{ + $\frac{77}{1000}$ = 50.0 ft. }	707.8	– –	+ 7.6	11249
58⅝	{ + 67.5 = 68.5 }	1075 5	– –	– 1.	11382
56	{ + 60. = 70.4 }	1244.4	– –	– –	11354
51¼	{ + 47.5 = 62. }	1365.5	– –	+ 1.	11254
53¼	{ + 52. = 63.2 }	1279.7	– –	– –	11396
52¼	{ + 51. = 101.5 }	2091.3	– –	– 6.7	11554
50½	{ + 43.5 = 136.7 }	3279.	– –	– 2.	11693
46⅛	{ + 34 3 = 107. }	3252.5	– –	–29.5	11690
49¼	{ + 42. = 32.6 }	810.	– –	– 8.8	11851
×2.1	{ + 22.7 = 71.3 }	3213.6	– –	–67.4	
52	{ + 74. = 25.4 }	368.3	– –	– –	11083
53	{ + 77 8 = 26.8 }	371.3			
51	{ + 46. = 15.8 }	359.2	364.4	+1.9	11182
44½	{ + 30.5 = 10.7 }	361 0			
44	{ + 29. = 10.2 }	362 0	– –	– –	10875
45¼	{ + 32.3 = 20.4 }	652.	– –	–2.0	10946

Level of the Clyde, and Stonebyre-hill, 654 feet.	Sept. 7. 8h 15′ A.M.	29.872 29.148	48½ 46¼	—055 —045	29.817 29.103	631.6	—22.4 =35.4	46¾ 45¼	45¼
Carmichael-well, and West-end of Carmichael-hill, 451½ feet.	July 30. 5h 40′ P.M. S.W. wind, beginning to rain.	29.162 28.690	56 54½	—076 —071	29.086 28.619	421.8	—29.7 =70.3	54¾ 53¼	54
	Aug. 1. 11h 40′ A.M.	29.612 29.135	58¾ 60	—086 —089	29.526 29.046	427.1	—24.4 =57.1	57 54¾	55⅞
Carmichael-well, and top of Tinto, four feet below the summit of the Cairn, 1642.5 ft.	June 30. 1h 30′ P.M.	28.991 27.284	61¾ 55¼	—093 —069	28.898 27.215	1563.6	—78.9 =50.5	58 51	54½
	July 30. 6h A.M.	29 063 27.335	51½ 46¼	—062 —043	29.001 27.292	1582.6	—59.9 =38.	51 44	47½
	Aug. 2. 8h 15′ A.M.	29.608 27.846	54½ 47½	—072 —048	29.536 27.798	1580.3	—62.2 =39 3	51½ 44½	48
	Aug. 27. 11h 50′ A.M. S.W. wind.	28.710 27.008	59¾ 53¼	—087 —063	28.623 26.945	1570.3	—72.2 =46.	55¾ 47¼	51½
	Aug. 27. 1h 40′ P.M. hail.	28.736 27.032	60¼ 53	—090 —062	28.646 26.970	1571.	—71.5 =45.5	55¼ 50	52¾
	Aug. 27. 1h 50′ P.M.	28.716 27.010	58½ 52¼	—083 —061	28.633 26.949	1579.5	—63. =40.	55½ 48¼	52
Mean of the observations on Tinto, with Mr. DE LUC's equation for the air, — — —						1574.5	51°—39°.7=11.3 × 2.1		

15¼	$\left\{\begin{array}{l}+\ 32.3\\ =\ 20.4\end{array}\right\}$	652.	– –	–2.0	10946
54	$\left\{\begin{array}{l}+\ 52.3\\ =\ 22.\end{array}\right\}$	443.8 ⎫	– –	–7.7 ⎫	
55⅝	$\left\{\begin{array}{l}+\ 57.1\\ =\ 24.4\end{array}\right\}$	451.5 ⎭	– –	– – ⎭	11430
54½	$\left\{\begin{array}{l}+\ 53.\\ =\ 83.3\end{array}\right\}$	1646 9 ⎫	– –	– –	11684
57½	$\left\{\begin{array}{l}+\ 36.4\\ =\ 57.4\end{array}\right\}$	1640.	– –	– –	11412
58	$\left\{\begin{array}{l}+\ 39.6\\ =\ 62.4\end{array}\right\}$	1642.7 ⎭			
51½	$\left\{\begin{array}{l}+\ 45.6\\ =\ 72.3\end{array}\right\}$	1642.6	⟩ 1645.5	+3.0	
52¼	$\left\{\begin{array}{l}+\ 49.\\ =\ 76.8\end{array}\right\}$	1647.8 ⟩	– –	– –	11704
52	$\left\{\begin{array}{l}+\ 46.4\\ =\ 73.3\end{array}\right\}$	1652.8 ⎭			
52.1	$\left\{\begin{array}{l}+\ 24.8\\ =\ 39.\end{array}\right\}$	1613.5	– –	–29.	1

N° IV. Computations of barometrical obſervations on heights n

Date of the obſervations, winds, &c.	Obſerved heights of the inferior and ſuperior barometers.	Temperature of the quickſilver.	Equation for the heat of the quickſilver.	Equated heights of the barometers.	Logarithmic reſult in feet.	Logar. exceſs or defect in ft. and alſo in 1000th parts.	Temperature of the air, Particular.	Temperature of the air, Mean.
1774, Aug. 12. 5h 20′ A.M.	30.086 29.704	52°½ 49¼	—067 —058	30.019 29.646	325.8	—18.2ft. = 56/1000	50°¾ 49	50°
Aug. 15. 6h 45′ A.M.	29.568 29.197	55¼ 53½	—075 —068	29.493 29.129	323.6	—20.4 =60.3	54 54½	54¼
Aug. 15. 0h 15′ P.M. S.W. wind and rain.	29.625 29.282	56¼ 53½	—078 —068	29.547 29.214	319.1	—24.9 =78.	54 54½	54¼
Aug. 15. 5h 15′ A.M. S.W. wind and rain.	29.567 28.704	55¼ 51¼	—075 —062	29.492 28.642	762.	—41. =53.8	54 50½	52¼
Sept. 15. 10h 30′ A.M. S.W. wind.	29.953 28.291	57½ 52¼	—084 —063	29.869 28.228	1472.5	—71.5 =48.6	54½ 47¼	51
Sept. 15. 1h 15′ P.M. S.W. wind.	29.561 28.272	63¾ 54	—100 —068	29.461 28.204	1136.2	—63.8 =56.2	56½ 48½	52½
1774, Dec. 1. 2h 45′ P.M.	29.565 28.770	35 32	—010 — —	29.555 28.770	701.5	— 0.9 = 1.3	33 30½	31¾
Dec. 10. 9h 46′ A.M.	29.494 28.687	20½ 20½	+038 +037	29.532 28.724	722.9	+20.5 =28.3	20½ 20½	20½
1775, Jan. 26. 1h 35′ P.M.	29.490 28.674	26½ 24¼	+018 +026	29.508 28.700	723.5	+21.1 =29.	26 23	24½
Nov. 10. 11h 30′ A.M.	29.959 29.177	38 34	—020 —006	29.939 29.171	677.2	— 6.8 =10.	36¾ 34	35½
Nov. 17. 9h 30′ A.M.	29.543 28.769	33¼ 30¼	—004 +005	29.539 28.774	683.8	—— ——	32 29¼	30⅝
1776, Jan. 31. 10h 45′ A.M.	30.009 29.229	15½ 24	+056 +026	30.065 29.225	711.7	+27.7 =39.	14 20	17
July 25. 2h 20′ P.M.	30.157 29 427	70¼ 66½	—125 —111	30.032 29.316	628.8	—55.2 =87.7	69¾ 67	68½
1775, Dec. 27. 11h 30′ A.M.	29.807 28.985	30¼ 29¾	+004 +007	29.811 28.992	725.9	— 4.9 = 6.7	29½ 29½	29½
Dec. 27. 8h 40′ A.M.	29.778 28.945	35¾ 33	—013 —003	29.765 28.942	730.6	—— ——	35¾ 32½	34⅛
1776, Feb. 1. 9h 30′ A.M.	29.883 29.032	28¾ 26½	+011 +019	29.894 29.051	745.4	+14.6 =19.6	24½ 26½	25½
Aug. 3. 2h 20′ P.M.	30.135 29.348	75½ 72	—141 —127	29.994 29.221	680.4	—50.4 =74.	72¾ 69	73¾

...erature [of th]e air, Mean.	Equation by the rule in 1000th parts, and also in feet.	Result by the rule, Particular.	Result by the rule, Mean.	Error of the rule in feet.	Ratio of the weight of quick-silver in air, ar...
50°	+ 47/1000 = 15.3f.	341.1	- -	- -	11037
54¼	+ 54. = 17.5	341.1	339.5	—4.5	
			- -	- -	11761
54¼	+ 54. = 17.2	336.3			
52¼	+ 50. = 38.	800.	- -	—3.	11309
51	+ 47. = 69.2	1541.7	- -	—2.3	11249
52½	+ 52. = 59.	1195.2	- -	—4.8	11410
31¾	— 0.5 = 0.3	701.2	- -	- -	10724
20½	—26.6 = 19.2	703.7	705.3	+2.9	11445
24½	—17. = 12.4	711.1	- -	- -	10419
35½	+ 8.6 = 5.8	683.	- -	- -	10646
30⅛	— 3.5 = 2.4	681.4	- -	- -	10729
			684.5	+0.5	
17	—35.2 = 25.	686.7	- -	- -	10184
68½	+ 94. = 58.2	687.0	- -	- -	11416
29½	— 5.7 = 4.1	721.8	- -	- -	10707
34¾	+ 5.4 = 4.	734.6	- -	- -	10642
			734.7	+3.9	
25½	+ 15.4 = 11.5	733.9	- -	- -	10390
73¾	+ 100. = 68.	748.4	- -	- -	11286

| Aug. 3. 2ʰ 20′ P.M. | 30.135
29.348 | 75½
72 | —141
—127 | 29.994
29.221 | | 680.4 | —50.4
=74. | 22
69 | 73¾ |

| In thefe two laft obfervations Mr. DE LUC's equation
for the air being fubftituted, | 1776, Feb. 1. | 745.4 | $39°.7—25°.5=14°.2 \times 2.1$ |
| | Aug. 3. | 680.4 | $70°.7—39°.7=31°. \times 2.1$ |

73¼	$\{$	+100. =68.	$\}$	748.4	| $-$ $-$	$-$ $-$	11286
×2.1	$\{$	−29.8 =22.2	$\}$	723.2	$-$ $-$	−7.6	
×2.1	$\{$	+65.1 =44 3	$\}$	724.7	$-$ $-$	−6.1	

N° V.

Nº V. Computations of barometrical obſervations made on heig
and Nº VI. of thoſe near Carnarvon in North Wal

Nº V. near Linhouſe.

Geometrical heights of the ſtations in feet.	Date of the obſervations, winds, &c.	Obſerved heights of the inferior and ſuperior barometers.	Temperature of the quickſilver.	Equation for the heat of the quickſilver.	Equated heights of the barometers.	Logarithmic reſult in feet.	Logar. exceſs or deficit in ft. and alſo in 1000th parts.	Temperature of the air, Particular.	Mean.	
Linhouſe and Eaſt-cairn hill, 5 feet below the ſummit, 1176.6 feet.	1775, Nov 11. 8ʰ A.M. calm and clear.	29.216 27.912	32° 30	—— +006	29.216 27.918	} 1184.2	{ +7.6 ft. = 6.4/1000	32° 29	} 30°¼	{ ⸗
	Nov. 15. noon.	28.941 27.632	32 27	—— +015	28.941 27.647	} 1191.9	{ + 15.3 = 12.8	32 26	} 29	{ ⸗
Linhouſe, and Eaſt-cairn hill, 18 feet below the top, 1165.6 feet.	1776, Dec. 17. 2ʰ P.M.	28.990 27.688	31¼ 24	+001 +025	28.991 27.713	} 1174.8	{ + 9.2 = 7.9	30¼ 22	} 26¼	{ ⸗
	Subſtituting Mr. DE LUC's equation for the air,			—		39°.7—26°.1 = 13°.6×2.1				{ ⸗
Linhouſe, and Weſt-cairn hill, 11 ft. below the top, 1178.4ft.	1775, Dec. 1. 1ʰ P.M. high S.W. wind, fog above.	29.250 28.003	49 45	—055 —042	29.195 27.961	} 1125.3	{ — 53.2 = 47.3	48° 45	} 46°¼	{ ⸗
Linhouſe, and Corſton hill, 4 feet below the top, 386.5 ft.	Dec. 8. 1ʰ P.M. clear and windy.	29.686 29.521	41 39	—029 —023	29.657 29.288	} 379.7	{ — 6.8 = 18.	40 39	} 39¼	{ ⸗
Corſton hill, and Weſt-cairn hill, 792 feet.	1776, Dec. 16ʰ 11′ A.M. high N. wind, clear weather.	28.580 27.714	34½ 32	—009 ——	28.571 27.714	} 793.6	{ + 1.6 = 2.	34½ 30	} 32⅚	{ ⸗
Corſton hill, and Eaſt cairn hill, 776.6 feet.	Dec. 17. 1ʰ A.M. light W. wind.	28.574 27.710	32 25	—— +022	28.574 27.732	} 779.4	{ + 2.8 = 3.6	31 23¾	} 27¼	{ ⸗
Linhouſe, and Corſton hill, 388.5 feet.	Nov. 20. 1ʰ P.M. ſnow had fallen, high W. wind.	27.992 27.582	35 33	—009 —003	27.983 27.579	} 379.	{ — 9.5 = 28.2	33 33	} 33	{ ⸗

Nº VI. near Carnarvon.

Carnarvon Quay, and Moel Eilio, 2371 feet.	1775, Aug. 4. 1ʰ 7′ P.M. rain above, clear below.	29.693 27.714	62½ 54	—098 —066	29.595 27.148	} 2248.8	{ —122.2 + 54.4	62½ 51	} 56½	{ ⸗
	Aug. 8. 0ʰ 7′ P.M. S. wind, and hazy weather above.	30.036 27.543	68 57	—118 —075	29.918 27.468	} 2226.3	{ —194.7 = 65.	68½ 56	} 62¼	{ ⸗
	Aug. 8. 2ʰ 7′ P.M. S. wind, weather ſomething clearer.	30.027 27.533	69½ 58¼	—122 —079	29.905 27.454	} 2228.3	{ —142.7 = 64.	69½ 57	} 63¼	{ ⸗
	Subſtituting Mr. DE LUC's equation for the air,				2231.1	60°.8—39°.7 = 21°.1×2.1				{ ⸗
	Aug. 7. 6ʰ 7′ A.M.	30.154 26.402	56½ 47½	—081 —045	30.073 26.417	} 3377.6	{ —177.4 = 52.5	56½ 45½	} 50¼	{ ⸗

heights near Linhouse.

h Wales.

rature air, Mean.	Equation by the rule in 1000th parts, and also in feet.	Result by the rule, Particular.	Result by the rule, Mean.	Error of the rule in feet.	Ratio of the weight of quick-silver to air, air being 1.
30°½	+ $\frac{3.4}{1000}$ = 4. ft.	1180.	1181.3	+ 4.7	10894
29	— 7. = 8.3	1182.6			
26⅛	— 14. = 16.4	1158.4	- -	— 7.2	10910
×2.1	= 28 6 = 33.6	1141.2	- -	—24.4	
46°½	+ 35. = 39.4	1164.7	- -	—13.7	11441
39½	+ 18. = 6.8	386.5	- -	- -	10736
32⅕	+ 0.3 = 0.2	793.8	- -	+ 1.8	11077
27¾	— 10.2 = 9.3	770.1	- -	— 6.5	11068
33	+ 2 2 = 0 8	379.8	- -	— 8.7	11540
6¾	+ 59.6 =134.	2382.8	- -	- -	11594
2⅜	+ 75. =167.	2393.3	2391.8	+20.8	
3¾	+ 76.8 =171.	2399.3	- -	- -	11566
×2.1	+ 44.3 = 98.8	- -	2330.	—41.	
0¾	+ 45. =152.6	3530 2	- -	- -	11646
3¾	+ 52.5 =177.4	3554.			

		[illegible]	[illegible]	[illegible]	[illegible]		[illegible]	[illegible]			–
	Aug. 7. 0ʰ 7′ P.M.	30.140 26.488	61½ 60½	—097 —083	30.043 26.405	3363.4	—191.6 = 57.	61¾ 54	57⅞		+ =
	Aug. 7. 2ʰ 7′ P.M.	30.144 26.478	62 53¾	—099 —063	30.045 26.415	3355.3	—199.7 = 59.5	62 51	56½		+ =
Carnarvon Quay, and Peak of Snowdon, 3555 feet.	Aug. 14. 8ʰ 7′ A.M. fog above.	29.984 26.271	56½ 42¾	—080 —031	29.904 26.240	3405.9	—149.1 = 43.8	55¼ 43	49⅛		+ =
	Aug. 14. 9ʰ 7′ fog and rain.	29.978 26.279	58½ 44	—087 —035	29.891 26.244	3390.6	—164.4 = 48.5	57¾ 43½	50¼		+ =
	Aug. 14. 10ʰ 7′.	29.972 26.280	60 44½	—091 —036	29.881 26.244	3381.9	—173.1 = 51.2	60 44½	52¾		+ =
	Aug. 14. 11ʰ 7′.	29.974 26.280	61½ 44¾	—097 —037	29.877 26.243	3379.4	—175.6 = 52.	61 45	53		+ =
	Aug. 14. 0ʰ 7′,	29.976 26.282	62½ 46½	—100 —042	29.876 26.240	3381.5	—173.5 = 51.3	62 46	54		+ =
	Barometrical height of Snowdon from the mean of two days observations,		—	—		3379.1	—175.9 = 52.1	— —	53.1		=
Mr. DE LUC's equation for the air,			—	—		53°.1—39°.7=13°.4 × 2.1=28.14					=

	(−177.4)			} 3551.		
7⅞	{ +61. / =205. }	3568.4	⎫		− − − −	11704
6½	{ +58.5 / =196. }	3551.3	⎭			
9⅜	{ +40. / =136.2 }	3542.1	⎫		− − − −	11643
0½	{ +44.3 / =150.4 }	3541.	⎭			
2¾	{ +48. / =162.7 }	3544.6	} 3546.8			
3	{ +50. / =169. }	3548.4	⎫		− − − −	11704
4	{ +52.5 / =176.2 }	3557.7	⎭			
3.1	{ =176.3	− −		3548.9	−6.1	
8.14	= 95.1	− −		3474.2	−80.8	

Compu-

Computations of part of Mr. DE LUC's barometrical obſervations, anſwering to the of the air.

	Stations with their geometrical heights in feet.	Date of the obſervations.	Obſerved heights of the inferior and ſuperior barometers.	Temperature of the quickſilver.	Equation for the heat of the quickſilver.	Equated heights of the barometers.	Logarithmic reſult in feet.	Logar. exceſs or defect in ft. and alſo in 1000th parts.	Temperature of the air, Particular.	Mean	Reduced.	Equation by the
Coldeſt of the Sun-riſing Obſervations.	1ſt. 230.5	1760, Feb. 9. 8h 30′ A.M.	28.986 28.703	33°¾ 26¼	−006 +018	28.980 28.721	233.9	+3.4 ft. = 14.5/1000	24¾° 26¼	25°½	—	— =
		March 9. 6h A.M.	28.875 28.586	37½ 28¾	−018 +012	28.857 28.598	234.9	+ 4.4 = 18.7	27½ 32	29¾	—	— =
	2d. 457.	March 9. 6h 8′ A.M.	28.875 28.342	37½ 30	−018 +006	28.857 28.348	463.7	+ 6.6 = 14.5	27½ 31½	29½	—	— =
	3d. 624.5	March 9. 6h 15′ A.M.	28.875 28.170	37½ 32	−018 ——	28.857 28.170	627.8	+ 4.3 = 7.	27¾ 34	30⅞	—	— =
	4th. 776.7	March 9. 6h 30′ A.M.	28.875 28.009	37½ 32	−018 ——	28.857 28.009	777.2	+ 0.5 = 0.7	27½ 32½	30	—	— =
	5th. 977.2	March 9. 6h 45′ A.M.	28.875 27.798	37½ 33¼	−018 −005	28.857 27.793	978.9	+ 1.7 = 1.7	27½ 32	29¾	—	— =
Coldeſt of the ordinary obſervations.	2d, 457.	Feb. 9. 9h A.M.	28.997 28.470	32 28½	—— +012	28.997 28.482	466.9	+ 9.8 = 21.	25¼ 29	27⅛	26°	— =
	3d, 624.5	Feb. 9. 9h 15′ A.M.	28.997 28.298	32 30	—— +006	28.997 28.304	630.3	+ 5.8 = 9.2	28 30	29	28	— =
	4th. 776.7	Feb. 9. 9h 30′ A.M.	28.997 28.142	32 32	—— ——	—— ——	780.2	+ 3.6 = 4.5	40½ 32	36¼	30½	— =
	5th. 977.2	Feb. 9. 10h A.M.	28.997 27.931	33¾ 35½	−006 −011	28.991 27.920	980.8	+ 3.6 = 3.5	40¼ 37	38⅝	31½	— =
	6th. 1298.9	Feb. 9. 10h 15′ A.M.	29.002 27.604	32 37½	—— −017	29.002 27.587	1303.4	+ 4.5 = 3.5	41 36¾	38⅞	31½	— =
	7th. 1513.3	Feb. 9. 10h 30′ A.M.	29.008 27.393	33¾ 41	−006 −027	29.002 27.366	1513.	− 0.3 = 0.2	41 37¾	39⅜	33½	+ =
	8th. 1938.9	Feb. 9. 11h A.M.	29.002 26.955	35½ 39¾	−011 −021	28.991 26.934	1917.7	− 21.2 = 11.0	43½ 37¾	40⅝	34½	+ =
	9th. 2094.5	Feb. 9. 11h 15′ A.M.	28.997 26.771	35½ 39¾	−011 −021	28.986 26.750	2091.8	− 2.7 = 1.5	43¾ 39¾	41¾	35	+ =
	10th. 2356.3	Feb. 9. 11h 45′ A.M.	28.992 26.494	35½ 39¾	−011 −021	28.981 26.473	2358.6	+ 2.3 = 1.	44¾ 38¾	41¾	35	+ =
	11th. 2486.3	Feb. 9. noon.	28.986 26.366	33¾ 37½	−006 −016	28.980 26.350	2479.	− 7.3 = 3.0	44¾ 36	40⅜	34½	+ =
	14th. 2922.	1759, July 15, 4h P.M.	28.759 25.950	74½ 71	−131 −105	28.628 25.845	2664.8	−257.9 = 96.7	88½ 74	81	75½	+1 =2

ed.	Equation by the rule in 1000th parts, and also in feet.	Result by the rule, Particular.	Mean.	Error of the rule in feet.	Ratio of the weight of quick-ſilver to air, air being 1.
·	$\{ \frac{-15.8}{1000} = 3\ \text{oit.} \}$	229.9	} 231.8	– –	10598
·	$\{ \frac{-4.9}{1.2} \}$	233.7		+1.3	10598
·	$\{ \frac{-56}{2.6} \}$	461.1	– –	+4.1	10732
·	$\{ \frac{-35}{2.3} \}$	625.5	– –	+1.	10876
·	$\{ \frac{-4.4}{3.4} \}$	773.8	– –	–2.9	10978
·	$\{ \frac{-5.}{4.9} \}$	974.0	– –	–3.2	11000
5°	$\{ \frac{-14.}{6.5} \}$	460.4	– –	+3.4	10649
3	$\{ \frac{-9.}{5.7} \}$	624.6	– –	– –	10814
2¼	$\{ \frac{-3.5}{a\ 7} \}$	777.5	– –	– –	10901
1¼	$\{ \frac{-1.1}{1.1} \}$	979.7	– –	+2.5	10949
1½	$\{ \frac{-1.1}{1.4} \}$	1302.	– –	+3.1	11024
3¼	$\{ \frac{+3.3}{5.} \}$	1518.	– –	+4.7	11120
4½	$\{ \frac{+6.}{11.5} \}$	1929.2	– –	+9.7	11306
5	$\{ \frac{+6.6}{12.6} \}$	2104.4	– –	+9.9	11241
5	$\{ \frac{+6.6}{15.6} \}$	2374.2	– –	+17.9	11274
4½	$\{ \frac{+5.6}{14.} \}$	2493.	– –	+6.7	11357
5¼	$\{ \frac{+101.}{=269.2} \}$	2934.	– –	+12.	12541

Hottest of the ordinary observations on the highest stations.	2922.	4 P.M.	25.930	71	—105	25.645		= [illegible]	74			=
	15th. 3119.2	July 15. 2ʰ P.M.	28.797 25.778	74½ 71	—131 —108	28.666 25.670	2876.5	—242.7 = 84.5	85 68¾	76¼	73½	+ =
		July 15. 3ʰ 30′ P.M.	28.764 25.778	74½ 68¾	—131 —103	28.633 25.675	2841.4	—277.8 = 97.7	90½ 74	82	76	+ =
	10th. 2356.3	1760, July 20. 10ʰ 15′ A.M.	28.775 26.499	71 72¾	—121 —117	28.654 26.382	2152.6	—203.7 = 94.7	87¼ 74	80⅝	75	+ =
		July 20. 3ʰ 15′ P.M.	28.731 26.460	74½ 72¾	—131 —118	28.600 26.342	2143.	—213.3 = 99.5	91½ 75¾	83⅝	77	+ =
	11th. 2486.3	July 20. 10ʰ 45′ A.M.	28.769 26.366	71 68¾	—121 —104	28.648 26.262	2266.	—221.3 = 97.7	87½ 74½	81	75	+ =
		July 20. 3ʰ P.M.	28.726 26.327	76½ 72¾	—138 —116	28.588 26.211	2262.	—224.3 = 99.	92¾ 76½	84¼	77	+ =
	12th. 2752.	July 20. 11ʰ 30′ A.M.	28.758 26.100	72¾ 68¾	—126 —103	28.632 25.997	2516.3	—235.7 = 94.	88 72½	80¼	74	+ =
		July 20. 2ʰ 30′ P.M.	28.720 26.066	76½ 71	—138 —110	28.582 25.956	2511.2	—240.8 = 95.7	92 77	84¼	77	+ =
	13th. 2877.5	July 20. noon.	28.747 25.977	72¾ 68¾	—126 —103	28.621 25.874	2629.3	—248.2 = 94.7	89 73½	81¼	75	+ =
		July 20. 1ʰ 45′ P.M.	28.720 25.691	74½ 68¾	—131 —105	28.859 25.856	2618.2	—259.3 = 99.	92¾ 75	83⅜	77	+ =
						Mean of the hottest,		95.7/1000	—	81.7	75.6	

3½	{ +96.5 / =300.8 }	3177 3	}	— —	— —	12439
				3154 4	+35.2	
6	{ +102. / =290 2 }	3131.6		— —	— —	12603
5	{ +102. / =219.6 }	2372.2	}			12429
				2371.2	+14.9	
7	{ +106. / =227.2 }	2370.2				
5	{ +101.8 / =230.7 }	2496.7	}			12468
				2499.2	+12.9	
7	{ +106. / =239.7 }	2501.7				
4	{ +97 6 / =245.6 }	2761.9	}			12504
				2769.6	+17.6	
7	{ +106. / =266.2 }	2777.4				
5	{ +101.8 / =267.6 }	2897.1	}			12548
				2896.4	+18 9	
7	{ +106. / =277.5 }	2895.7				

5.6

Compu-

ſtations with their geometrical heights in feet.	Date of the obſervations.	Obſerved heights of the inferior and ſuperior barometers.	Temperature of the quickſilver.	Equation for the heat of the quickſilver.	Equated heights of the barometers.	Logarithmic reſult in feet.	Logar exceſs or defect in ft. and alſo in 1000th parts.	Temperature of the air, Particular.	Mean	Reduced.	Equation by the rule in 1000th
The Dole, by Sir GEORGE SHUCKBURGH's meaſurement, is above the lake of Geneva 293 ft. Mr. DE LUC's lowermoſt barometer was higher than the lake 83 ft. hence the vertical diſtance of the barometers, 4210 feet.	1764, July 29. 1h P.M.	28.953 24.951	78°¾ 67	—145 - 095	28.808 24.856	3845.	—365 ft. = 94.9/1000	77¼° 64½	71°¾	70°	+ 88/10… =339
	July 29. 1h 30′ P.M.	28.942 24.940	78¼ 65¼	—145 —090	28.797 24.850	3841.	—369. = 96.	78¼ 65	71½	70	+ 8… =33…
	1765, July 21. 10h 30′ A.M.	28.698 24.640	67 59¼	—108 —075	28.590 24.565	3954.	—256. = 64.7	67½ 51	59¼	58	+ 6… =23…
	1764, July 8. 8h A.M.	28.692 24.636	71 57¼	—121 —070	28.571 24.566	3935.	—275. = 70.	73 56	64½	63	+ 7… =28.
Light-houſe of Genoa, 237.6 ft	1757, June 2. 6h A.M.	30.077 29.817	76 76	—142 —141	29.935 29.686	217.7	— 20. = 92.	— —	75¼	73	+10. = 2…
	June 2. 4h½ P.M.	30.088 29.846	82½ 84½	—163 —169	29.925 29.677	216.8	— 21. = 96.	— —	82	79	+10… = 2.
	June 23. 9h½ A.M.	30.116 29.857	79 75	—152 —138	29.964 29.719	214.8	— 22.8 =107.	— —	79	76	+10. = 2
	June 23. 5h 45′ P.M.	30.041 29.796	79 79	—152 —150	29.889 29.646	212.7	— 24.9 =117.	— —	78	75	+10 = 2
	July 26. 1h P.M.	30.021 29.774	83½ 83¼	—166 —164	29.855 29.610	214.7	— 22.9 =107.	— —	81	77	+11 = 2
								Mean	76		
For the barometrical height of Turin above Genoa,		30.019 +.071 30.090	77	—146	29.944	722.6	——	—	77	—	+11
		29.197 +.069 29.266	77	—141	29.125		——	—			= 8…
For the barometrical height of Mr. DE LUC's room above Turin,		29 319 28.831	72¾ 72¾	—129 —126	29.190 28.705	436.6	—— ——	— —	72¾	—	+ 9… = 4.

Mr. DE LUC's room above Genoa, — —
Surface of the Lake of Geneva in ſummer below Mr. DE LUC's room,

Surface of the Lake of Geneva above the Mediterranean, —

By Mr. DE LUC's rule the Lake is elevated above the Sea 1126 French, or 12…

In the obſervations on the Dole, if Mr. DE LUC's equation for the air is ſubſtituted inſtead of that reſulting from the Britiſh obſervations, the barometrical height will be, } 3894 66°.6—39°.7=

| Equation by the rule in 1000th parts, and also in feet. | Refult by the rule, | | Error of the rule in feet. | Ratio of the weight of quick-filver to air, air being 1. |
	Particular.	Mean.		
$\{$ + $\frac{88\,2}{1000}$ =339.8f. $\}$	4184 8		– – –	12714
$\{$ + 88 2 =338.8 $\}$	4179.8			
		4194.	—16.	
$\{$ + 60. =237. $\}$	4191.		– – –	12520
$\{$ + 72.4 =285. $\}$	4220.		– – –	12551
			Mean	12595
$\{$ + 103. = 22.4 $\}$	240.1			
$\{$ + 108. = 23.4 $\}$	240.2			
$\{$ + 104. = 22.3 $\}$	237.1	238.2	+ 0.6	12672
$\{$ + 106.5 = 22.6 $\}$	235.3			
$\{$ +111. = 23 8 $\}$	238.5			
$\{$ +111.5 = 80.6 $\}$	803.2			
$\{$ + 99. = 43.2 $\}$	479.8			

$$1283.$$
room, $\quad$ 53.3
— $\quad$ 1229.7

, or 1200 Englifh feet.

$-39°.7 = 26°.9 \times 2.1 \left\{ \begin{matrix} +\ 56.5 \\ =220. \end{matrix} \right\} = 4114.\ \overset{\text{Erroi}}{-96.0}$

Eaft fignal on the able-hill above the fea, —	3417	11ʰ 30' A.M.	30.174 / —.173 / 30.001	66—111 29.890		3192.	—211.	66°	58	+ 6
Height of the obfervatory, —	14									
Vertical diftance of the barometers in feet,	3403		26.502	52—058 26.444			= 66.	50		=19
Veft fignal above the fea, —	3468	0ʰ 30' P.M.	30.174 / —.197 / 29.977	66—111 29.866		3289.	—179.	66	58	+ 6
eight of the obfervatory, —	14									
rtical diftance the barome-rs, —	3454		26.377	50—053 26.324			=544.	50		=20
ith Mr. DE 's Equation he air,	Mean 3426	— 3240	58°—39°.7=18°.3×2.1 { +38.4 =124.4 }	3364.4	-	-	-	-	-	

$\left\{\begin{array}{l}+\ 62.4 \\ \\ =199.2\end{array}\right\}$ 3391.2	– –	–11.8	11713
$\left\{\begin{array}{l}+\ 62.4 \\ \\ =205.2\end{array}\right\}$ 3494.2	– –	+40.2	11662
		Mean	11687
– – – – – –		–62.6	

Compu-

Computations of barometrical obſervations made on heights that have not bee

geometrically.

Date.	Stations of the barometers.	Obſerved heights of the inferior and ſuperior barometers.	Temperature of the quickſilver.	Equation for the heat of the quickſilver.	Equated heights of the barometers.	Logarithmic reſult in feet.	Temperature of the air, Particular.	Mean.	Equation for the heat of the air
	Level of the ſea at Invergourie, and Belmont-caſtle.	29.932 29.734	54° 57	—072 —081	29.860 29.653	181.3	54° 54	54°	+ [illegible] =9.8
1773, July 8.	Superior barometer, Top of Kinpurney-hill.	29.988 28.974	65 62	—108 —095	29 880 28.879	887.9	63 57	60	+71. =63.
1776, Sept. 12.	Ditto.	30.331 29.275	56¼ 51	—080 —061	30.251 29.214	908.9	57½ 50	53¾	+56. =50.
	Caſtle Menziés.	29.756 29.674	60¼ 64½	—092 —104	29 .64 29.570	82.7	61 63	62	+74. = 6.
Sept. 11.	Top of Farragan.	29.794 27.344	63¾ 52½	—102 —062	29.692 27.282	2205.8	65 50	57½	+62. =137
Sept. 17.	Top of Ben Lawers.	29.800 25.830	55 38	—075 —017	29.725 25.813	3677.	54 36	45	+30. =11(
	Top of Ben More.	30.000 26.148	55½ 42	—077 —029	29.923 26.119	3542.9	52½ 37	44¾	+31. =109
Sept. 12.	Top of Ben Gloe.	29.712 26.142	62 48	—097 —041	29.615 26.101	3291.3	62 45	53½	+51. =167
Sept. 13.	Blair of Athol-lawn.	29.636 29.380	60 58	—091 —083	29.545 29.297	219.6	58 60½	59¼	+67. =14.
Aug. 22.	Top of King's Seat.	29.904 28.791	68 66½	—116 —108	29.788 28.683	985.	67 64	65½	+84. =83.
1775, Sept. 5.	Hill of Barry.	29.870 29.345	62 56	—098 —076	29.772 29.269	444.	60 56	58	+64. =28.
Sept. 5.	Dunſinane-hill,	29.784 28.913	62 59	—097 —086	29.687 28.827	766.	62 59	60½	+71. =54
1774, Aug. 29 and 30. mean of three obſervations.	Quay at the new bridge of Glaſgow, and ſtation at Lanark.	29.560 28.850	55¾ 52¼	—077 —066	29.483 28.784	625.2	53¾ 50¼	52¼	+50. =31.
1772, Aug. 6. 2^h P.M.	Freeport in the iſland of Iſla, 19 feet above the ſea, and ſummit of the South-pap of Jura,	30.224 27.642	67 57	—114 —076	30.110 27.566	2300.2	60 57	58½	+66. =152
1772, Sept. 25. 9^h 25′ A.M.	Hafniford in Iceland, at the ſea-ſhore, and ſummit of Mount Hecla.	29.859 24.722	49 38	—056 —016	29.803 24.706	4886.8	43 24	33½	+ 3 =16.

The inferior barometer at Belmont-caſtle.

ot been determined

n.	Equation for the heat of the air in 1000th parts, and in feet.	Vertical distance of the barometers.	Horizontal distance of the barometer in miles.
	$+\frac{54.2}{1000}$ = 9.8 f.	191.1	10¼
	+ 71. = 63.1	951.	2⅛
	+ 56. = 50.8	955.3 959.7	2⅝
	+ 74. = 6.1	88.8	28½
	+ 62.4 = 137.8	2343.6	29
	+ 30. = 110.	3787.	42
	+ 31. = 109.8	3652.7	53½
	+ 51. = 167.9	3459.2	27¼
	+ 67. = 14.7	234.3	30
	+ 84.3 = 83.	1068.	6¼
	+ 64. = 28.4	472.4	4½
	+ 71.5 = 54 8	820.8	7½
	+ 50.0 = 31.3	656.5	22½
	+ 66.3 = 152.5	2452.7	4¼
	+ 3 4 = 16.6	4903 4	76

mit of Mount Hecla.

Compu-

Computations of Mr. BOUGUER's obfervations in Peru, fuppofing them to have be[
mean temperature of the day, between the coldeft of the morning [

Relative heights of the ftations, with refpect to the South-fea,		Stations of the barometers, with their geometrical diftance in feet.		Obferved heights of the inferior and fuperior barometers.	Temperature of the quickfilver.	Equation for the heat of the quickfilver.	Equated heights of the barometers.	Mean heights of the barometers.	Logarithmic refult in feet.	Logar. excefs or defect in
Heights of the columns of air, whofe bafes ftood at the fea,		15833	{ South-fea, Coraçon,	29.930 / 16.808	84½° / 43½	—169 / —022	29.761 / 16.786	23.27	14922.	{ = 9
		15564	{ South-fea, Pichincha,	—— / 16.963	84½ / 44½	—— / —024	29.761 / 16.939	23.35	14685.6	{ —8? / = 6
		9374	{ South-fea, Quito,	—— / 21.403	84½ / 65½	—078	29.761 / 21.325	25.54	8685.5	{ —68 / = 8
		7840	{ South-fea, Carabourou,	—— / 22.625	84½ / 66½	—084	29.761 / 22.541	26.15	7240.5	{ —59 / = 8
Superior fections of the columns of air, with the diftances of their bafes from the fea,	Coraçon, 15833 / Carabourou, 7840 }	7993	{ Carabourou, Coraçon,	——	66½ / 43½	——	22.541 / 16.786	19.66	7681.6	{ —31 / = 4
	Pichincha, 15564 / Carabourou, 7840 }	7724	{ Carabourou, Pichincha,	——	66½ / 44½	——	22.541 / 16.939	19.74	7445.1	{ —27 / = 3
	Coraçon, 15833 / Quito, 9374 }	6459	{ Quito, Coraçon,	——	65½ / 43½	——	21.325 / 16.786	19.05	6236.5	{ —22 / = 3
	Pichincha, 15564 / Quito, 9374 }	6190	{ Quito, Pichincha,	——	65½ / 44½	——	21.325 / 16.939	19.13	6000.1	{ —18 / = 3
			Mean of the four fuperior columns,	- -	- -	- -	- -	- -	- -	
	Quito, 9374 / Carabourou, 7840 }	1534	{ Carabourou, Quito,	——	66½ / 65½	——	22.541 / 21.325	21.93	1445.	{ —8 / = 6

ave been made at correſponding times, and in the
ming and hotteſt of the afternoon.

Logar. exceſs or defect in ft. and alſo in 1000th parts.	Mean temperature of the air.	Equation for the heat of the air in 1000th parts, and in feet.	Reſult by the rule in feet.	Error of the rule.	Ratio of the weight of quick-ſilver to air, Particular	Mean.
{− 911 ft. / = $\frac{61}{1000}$}	64°	{+ $\frac{61.6}{1000}$ / =919. ft.}	}15841.	+ 8.0	14590 ⌉	
{−878.4 / = 60.}	64¼	{+ 62. / =920.4}	}15606.	+42.	14517 ⌋	14553
{−688.5 / = 80.}	75	{+ 90. / =781.7}	}9467.2	+93.2	13273 ⌉	
{−599.5 / = 83.}	75½	{+ 96.5 / =698.7}	}7939.2	+99.2	12968 ⌋	13120
{−311.4 / = 40.5}	55	{+ 35.2 / =274.4}	}7952.	−41.	16623 ⌉	
{−278.9 / = 37.3}	55¼	{+ 36.5 / =271.7}	}7716.8	− 7.2	16507 ⌋	16565
{−222.5 / = 35.7}	54¼	{+ 33.2 / =207.}	}6443.5	−15.5	17149 ⌉	
{−189.9 / = 31.6}	55	{+ 34. / =204.}	}6204.1	+14.1	16893 ⌋	17021
$\frac{36.3}{1000}$	55	− − −	− −	− −	16793	
{− 89. / = 61.6}	66	{+ 61. / = 88.}	}1533.	− 1.	15089	

ROMETRICAL OPERATIONS.

P L A N

OF THE

IANGLES

ute ufe of for obtaining

etrical Diftance and Altitude

WDON and MOEL EILIO

with refpect to the Sea at

ARNARVON.

Aug.ᵗ 1775.

Scale of Feet.

...the length of the iron chain made use of on this occasion, was ascertained by means of accurate Deal Rods, applied to it every Morning & Evening, before and after the operation of the Field. The two measurements agreed to within less than a Foot, and made the length of the Base — 14076 F. This Base AB, was afterwards prolonged to C, the Top of the Dinas by means of the side Base BF 411.7 feet. Thus the distance BC, being 482.2 feet; the total Base AC amounts to 14558.2 feet.

N, High Watermark Neap Tide. D, a small Eminence called the Diged, T, the Foot Hill of Carnarvon. S, The Peak of Snowdon. E, A Cairn of Stones on Moel Eilio.

The Latitude of Carnarvon was found to be 53.9.47; and the Variation of the Needle, by two Azimuths of the Sun, taken on the 13th of Augt at 8 F. and 3 h. P.M. 21.18.30 westerly.

The Height of Moel Eilio above Carnarvon Quay 2371 Feet, resulting from the simplest operation on the Diged, the nearest point to the Hill, is to be preferred to that deduced from the Angles of Elevation taken from A.

B C, and distance of N from C.	Obs.d	NBC	98.46.50	NC....612.5
		BCN	30.8.20	
		CNB	51.4.50	
ACD and DBC, for the distance of D from AB and C; and for the Angles CBD and BDC.	Obs.d	DAC	123.17.50	AD..17269.7
	Cont.d	ADC	25.48.30	CD..27940.7
		ACD	30.53.40	BC...482.2
		DCB	30.53.40	BD..27557.4
		CBD	148.35.24	
		BDC	0.30.56	
	Obs.d	ABS	92.38.15	
	—	ABD	31.34.36	
DSB for the difference of Snowdon from B and D.	Obs.d	DBS	61.11.30	BS..57366.2
	=	SDC	69.36.50	DS..50258.3
	+	CDB	0.30.56	
	—	SDB	90.6.40	
	Hence	DSB	28.41.35	

NB. The time not admitting of any actual Survey of the Envi... to be consider'd as a slight Sketch intended merely to convey a g...

	DEB and DAE,	Obs.ᵈ	EDC	88. 47 . 6	DE .. 29349. 6		S above B	3360. 1	of Moel Eilio
1	For the diſtance of	+	CDB	0. 30 . 36		For the	C above B Trigᵗ 97.3		above D, and
	Moel Eilio from		EDB	86. 58 . 1		Altitude of	Inſtrument 4. 6		the Sea at
7	D and A .		DEB	44. 37 . 23		Snowdon	C above B 101. 9		Carnarvon .
7		Conſt	ADE	32. 32 . 30		above B and N	C above N Trigᵗ 114. 3		
2			DAE	43. 49 . 30	AL . 29190. 2		Inſtrument 4. 6		
			DEA	13. 50 . 14			C above N 109. 7		
4	CAT, for the	Obſ.ᵈ	CAT	123. 9 . 17			Hence B above N	7. 6	
	diſtance of the		ACT	29. 36 . 45	AT . 15430		Snowdon above the Sea at N	3567. 9	
	Toot Hill from A.		ATC	27. 33 . 55			S above D Trigᵗ	3558. 7	For the Altitude
			C	11 . 24 . 37			Curvature & Refraction ... 47. 3		of Moel Eilio
3	Angles	From..B...	S	3 . 49 . 7		For the	Inſtrument 4. 6		above A, and
3	of Elevation	From..C Dep.	N	10. 34 . 23		Altitude of	S above D	3509. 1	the Sea at
	and	From ..D...	A	3. 47. 54		Snowdon	D above North Angle of		Carnarvon .
	Depreſsion		E	4. 45 . 35		above D and the	Carnarvon Quay by Levelⁿ	106. 5	
		From ..A...	E	3 . 24 . 17		Sea at Carnarvon	Vertical diſtance of the Barometer	3555. 4	
			T	0 . 30. 47			Quay above High Water N.Tide	13.	
							Snowdon above the Sea	3568. 4	

he Environs of Carnarvon & Snowdon to be made : the Plan is only
vey a general Idea of the nature of the Country where the Triangles were ſituated.

Table, shewing the equation to be applied to the observed height of quicksilver in the barometer, from 15 to 32 inches, and for difference of temperature extending to 102° of FAHRENHEIT, whereby the column is reduced to the height it would have stood at in the temperature of 32°.

TABLE II. Results of experiments on the expansion of air of the density of five-tenths of the common atmosphere, and of others on air that was extremely rare, being only possessed with about one-tenth of an atmosphere.

TABLE IV. Results of experiments on the expansion of air, gradually condensed, by the admission of steam, and sometimes water, into the bulb of the thermometer.

Table shewing the expansion depending on the temperature of the column of air, and its elevation above the fire, as deduced by the mean height of quicksilver in the sub-
near and superior barometers.

Nº I. Computations of barometrical observations made on heights in and near London.

No. II. Comparison of barometrical observations made on heights near Tynbridge in Perthshire; and No. III of those near Lanark.

| No. II. near Tynbridge. |

Nº IV. Composition of barometrical observations on heights near Edinburgh.

Nº V. Comparisons of barometrical observations made on heights near Lusborde, and Nº VI of those near Carnarvon in North Wales.

Nº V. near Lusborde

Nº VI. near Carnarvon.

Compa-

Computations of part of Mr ☐☐ ☐☐☐'s barometrical observations, referring to the coldest and hottest temperatures of the air.

Continuation of Mr De Luc's barometrical observations.

Mr. de la Caille's barometrical Observations, Sept. 6th, 1751, at the Cape of Good Hope

Comparison of barometrical observations made on heights that have not been determined geometrically.

Computations of Mr. nairne's observations at Paris, supposing them to have been made at corresponding times, and in the mean temperature of the day, between the orbits of the morning and hours of the afternoon.

PLAN
OF THE
TRIANGLES